T0308734

Principles of Biophotonics, Volume 3

Field propagation in linear, homogeneous, dispersionless, isotropic media

Online at: https://doi.org/10.1088/978-0-7503-1646-0

IPEM–IOP Series in Physics and Engineering in Medicine and Biology

About the Series

The series in Physics and Engineering in Medicine and Biology will allow the Institute of Physics and Engineering in Medicine (IPEM) to enhance its mission to 'advance physics and engineering applied to medicine and biology for the public good'.

It is focused on key areas including, but not limited to:
- clinical engineering
- diagnostic radiology
- informatics and computing
- magnetic resonance imaging
- nuclear medicine
- physiological measurement
- radiation protection
- radiotherapy
- rehabilitation engineering
- ultrasound and non-ionising radiation.

A number of IPEM–IOP titles are being published as part of the EUTEMPE Network Series for Medical Physics Experts.

A full list of titles published in this series can be found here: https://iopscience.iop.org/bookListInfo/physics-engineering-medicine-biology-series.

Principles of Biophotonics, Volume 3

Field propagation in linear, homogeneous, dispersionless, isotropic media

Gabriel Popescu

Department of Electrical and Computer Engineering, Beckman Institute for Advanced Science and Technology, University of Illinois at Urbana–Champaign, Illinois, USA

IOP Publishing, Bristol, UK

ISBN   978-0-7503-1646-0 (ebook)
ISBN   978-0-7503-1645-3 (print)
ISBN   978-0-7503-1954-6 (myPrint)
ISBN   978-0-7503-1647-7 (mobi)

DOI   10.1088/978-0-7503-1646-0

Version: 20221201

IOP ebooks

British Library Cataloguing-in-Publication Data: A catalogue record for this book is available
from the British Library.

Published by IOP Publishing, wholly owned by The Institute of Physics, London

IOP Publishing, No.2 The Distillery, Glassfields, Avon Street, Bristol, BS2 0GR, UK

US Office: IOP Publishing, Inc., 190 North Independence Mall West, Suite 601, Philadelphia,
PA 19106, USA

Contents

Acknowledgement

I am grateful to my teachers, colleagues, and students, from whom I have been learning every day. In preparation of the manuscript, I received generous support from the ECE Department at UIUC.

I would like to acknowledge help with typesetting and figures from Ionut Preoteasa, and with proof reading by Chenfei Hu and Michael Fanous. Finally, I am grateful to IOP Publishing for their assistance.

Author biography

Gabriel Popescu

Gabriel Popescu (deceased) was a Professor in Electrical and Computer Engineering, University of Illinois at Urbana-Champaign. He received his PhD in Optics in 2002 from the School of Optics/CREOL (now the College of Optics and Photonics), University of Central Florida and continued his training with Michael Feld at M.I.T., working as a postdoctoral associate. He joined Illinois in August 2007, where he directed the Quantitative Light Imaging Laboratory (QLI Lab) at the Beckman Institute for Advanced Science and Technology. Dr Popescu served as Associate Editor of the journals *Optics Express* and *Biomedical Optics Express*, and as an Editorial Board Member for *Journal of Biomedical Optics and Scientific Reports*. He was an OSA and SPIE Fellow.

IOP Publishing

Principles of Biophotonics, Volume 3
Field propagation in linear, homogeneous, dispersionless, isotropic media
Gabriel Popescu

Chapter 1

Maxwell's equation in integral form

1.1 Faraday's law

Faraday is responsible for ground-breaking experiments in electromagnetism (see reference [1] for his 1832 treatise). The equation that bears his name states that a time-changing magnetic flux generates a circulating electric field and vice-versa, namely

$$\oint_C \mathbf{E} \cdot \mathrm{dl} = -\frac{\mathrm{d}}{\mathrm{d}t} \int_S \mathbf{B} \cdot \hat{\mathbf{n}} \; da. \tag{1.1}$$

In equation (1.1), $\mathbf{E}$ is the induced electric field (in units of V m^{-1}), $\mathbf{B}$ is the magnetic flux density, or magnetic inductance (in units of Tesla, or kg s^{-1} A^{-1}), the left-hand side integral is along a closed path, while the right-hand side is over an area S.

The integral on the right-hand side denotes the magnetic flux,

$$\phi_B = \int_S \hat{\mathbf{B}} \cdot \hat{\mathbf{n}} \; da, \tag{1.2}$$

where $\hat{n}$ is the unit vector defining the normal to the area element da.

The integral on the left-hand side of equation (1.1), sometimes called the *electromotive force (emf)*, represents the *circulation* of vector $\mathbf{E}$, which can be expressed in terms of the electric force, $\mathbf{F} = q\mathbf{E}$, with q the electric charge,

$$\begin{aligned} \int_C \mathbf{E} \cdot \mathrm{dl} &= \frac{1}{q} \oint_C \mathbf{F} \cdot d\mathbf{l} \\ &= \frac{W}{q}. \end{aligned} \tag{1.3}$$

In equation (1.3), W represents the work performed by the electric force, or, the energy transferred to the charge q. Thus, from equations (1.1)–(1.3), we can conclude that the change in the magnetic flux is equal to the energy per unit of charge,

doi:10.1088/978-0-7503-1646-0ch1

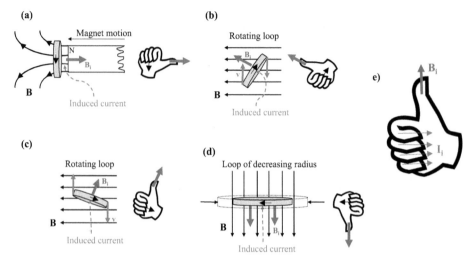

Figure 1.1. The induced current and induced magnetic field in several situations, and Lenz's law, indicated by the right-hand rule in each case. (a) Magnet moving toward the loop tends to increase the magnetic flux, thus, the induced current in the loop is such that it generates an induced magnetic field $\mathbf{B_i}$ in the opposite direction to $\mathbf{B}$, to oppose the flux increase. (b) Rotating a loop in the magnetic field $\mathbf{B}$ as shown by velocity $\mathbf{v}$ tends to reduce the magnetic flux, thus, the induced current in the loop is such that it generates and induced magnetic field $\mathbf{B_i}$ that adds to $\mathbf{B}$ and opposes the flux decrease. (c) As the loop passes the horizontal plane, the signs of the induced current and magnetic field switch, as now the rotation causes the magnetic flux to increase. (d) Reducing the size of a loop in a magnetic field tends to decrease the magnetic flux, thus, the induced current is such that the induced magnetic field $\mathbf{B_i}$ adds to the original field $\mathbf{B}$ and opposes the flux decrease. (e) Right-hand rule: if the curled fingers point in the direction of the rotating induced current, I_i, thumb points in the direction of the induced magnetic field, $\mathbf{B_i}$.

$$\frac{W}{q} = -\frac{d\phi_B}{dt}.$$
(1.4)

According to the definition in equation (1.2), the magnetic flux can change due to several different factors: (i) change in magnetic field, $\mathbf{B}$, (ii) change in the orientation of the surface normal, $\hat{\mathbf{n}}$, or (iii) change in the absolute value of the area, S. These situations are illustrated in figure 1.1. The negative sign in Faraday's law is significant: it establishes that the induced current has such a sign as to generate a magnetic field that opposes the change in magnetic flux. This result is known as *Lenz's law*, and can be used to explain the situations in figure 1.1, as described in the figure caption. Note that the 'right-hand rule' is a convenient way to find the directions of the induced current and magnetic field: if the curled fingers align with the direction of the current, then the thumb points in the direction of the magnetic field. In order to find out the direction of the induced current in the loop, we first find the *induced* magnetic field that opposes the change in magnetic flux, align our right thumb with it, then find the induced current following our curled fingers. Figure 1.2 shows other examples of applying Lenz's law, which involve a permanent magnet and a conducting loop. The reader can verify that the right-hand rule is fulfilled.

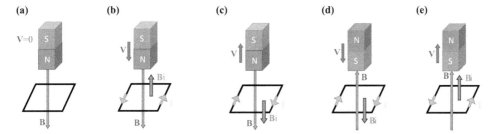

Figure 1.2. Induced currents I (green arrows) and magnetic fields $\mathbf{B_i}$ (orange arrows), when a permanent magnet **B** (north and south poles as indicated) is static (a), moves toward the loop with N-pole facing the loop (b), moves away from the loop with N-pole facing the loop (c), moves toward the loop with S-pole facing the loop (d), moves away from the loop with S-pole facing the loop (e).

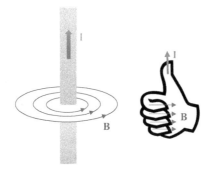

Figure 1.3. Ampère's law: a constant current I generates a rotating magnetic field **B** according to the right-hand rule.

Next, we discuss the situation when the magnetic field circulates in a loop and the electric field flux varies in time. The relationship between the two quantities is called Ampère's law and constitutes the second Maxwell equation.

1.2 Ampère's law

Ampère established a relationship between a *constant* electric current and a *circulating* magnetic field. Maxwell added another term, a *varying electric flux*, and showed that it, too, can generate a circulating magnetic field. The resulting equation, more accurately described as the Ampère–Maxwell equation, reads

$$\oint_C \mathbf{B} \cdot d\mathbf{l} = \mu_0 \left(I + \varepsilon_0 \frac{d}{dt} \int_S \mathbf{E} \cdot \hat{\mathbf{n}} \, da \right) \tag{1.5}$$

In equation (1.5), **B** is the induced magnetic field (in units of Tesla) around a closed path, C. On the right-hand side, I is the electric current, and the last term describes a varying electric flux. The parameters μ_0 and ε_0 are the magnetic permeability and electric permittivity of free space, respectively, $\mu_0 = 4\pi \ 10^{-7}$ H/m (Henry per meter), $\varepsilon_0 = 8. \ 85 \times 10^{-12}$ F/m (Farads m^{-1}). Equation (1.5) states that a changing electric flux or an electric current through a surface generate a circulating magnetic field around that surface and vice versa. Figure 1.3 illustrates the right-hand rule for

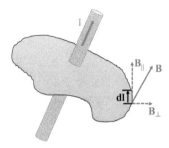

Figure 1.4. Ampère's law on an arbitrarily shaped loop: only $\mathbf{B}_{\parallel}$ contributes to the closed loop integral, as it contains the **B·dl**, with **dl**, the infinitesimal loop path element.

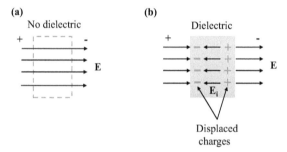

Figure 1.5. Electric field induced in a dielectric. (a) No dielectric present: the applied electric field **E** is unchanged. Note that **E** points from + to −. (b) Dielectric present: the applied electric field **E** displaces charges inside the material, such that the induced field $\mathbf{E}_i$ points in a direction opposite to **E**.

Ampère's law, where the thumb points to the current and curled fingers to the rotating magnetic field. Figure 1.4 describes the general situation of an arbitrarily shaped loop, emphasizing that only the tangential component of **B** contributes to the integral on the left-hand side in equation (1.5).

The Ampère–Maxwell law applies to materials other than free space. In this case, μ_0 and ε_0 must be replaced with μ and ε describing the magnetic and electric properties of the particular medium. The relative constants are defined with respect to the vacuum values, namely

$$\varepsilon_r = \varepsilon/\varepsilon_0 \tag{1.6a}$$

$$\mu_r = \mu/\mu_0 \tag{1.6b}$$

The dielectric permittivity (or constant) accounts for the fact that the electric field inside a dielectric material is generally different from the applied electric field. This phenomenon can be understood at the microscopic scale by noting that the electronic charge is displaced by the external field, as shown in figure 1.5. The charge displacement is such that it creates an internal electric field that has the opposite sign with respect to the external field. Thus, the total field inside the material is lower than the external field.

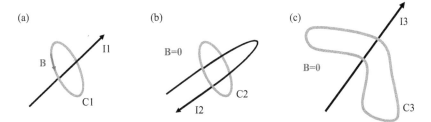

Figure 1.6. (a) Current enclosed by the loop. Currents not enclosed by the loop are shown in (b) and (c).

The magnetic permeability similarly accounts for the fact that, inside the material, the magnetic field can be different from the applied external field. Depending on how μ_r compares to unity, magnetic materials are classified as *diamagnetic*, *paramagnetic* and *ferroelectric*. Paramagnetics are characterized by $\mu_r > 1$. For example, aluminum has $\mu_r = 1 + 2 \times 10^{-5}$. Diamagnetic materials have $\mu_r < 1$. For example, gold and silver have, approximatively, $\mu_r = 1 - 3 \times 10^{-5}$. Ferromagnetism is a much stronger effect ($\mu_r = 5000$ for pure iron), which is responsible for permanent magnets. For these materials, the permeability depends on the applied magnetic field.

It is important to remember that the current I in equation (1.5) is enclosed by the path C, over which the magnetic field circulates. Figure 1.6 illustrates currents that are and are not enclosed by the path. Thus, figure 1.6(a) shows an enclosed current, while figures 1.6(b) and (c) show two distinct cases where the net current enclosed by the path is zero. Note that for the case in figure 1.6(a) the right-hand side applies: while the thumb points in the direction of the current, the fingers curl in the direction of the magnetic field.

The change in the electric flux can be expressed in an insightful way, using the following, identities

$$\int_S \mathbf{E} \cdot \hat{\mathbf{n}} \, da = \int_S \frac{\sigma}{\varepsilon_0} da$$
$$= \frac{Q}{A\varepsilon_0} \int_S da \qquad (1.7)$$
$$= \frac{Q}{\varepsilon_0}.$$

In equation (1.7), σ is the charge density, Q the charge and A transverse area (figure 1.7). Thus, the corresponding term in Ampère's law can be written as

$$\varepsilon_0 \frac{d}{dt} \int_S \mathbf{E} \cdot \hat{\mathbf{n}} \, dA = \varepsilon_0 \frac{d}{dt}\left(\frac{Q}{\varepsilon_0}\right)$$
$$= \frac{dQ}{dt} \qquad (1.8)$$
$$= I_d.$$

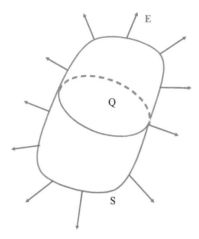

Figure 1.7. Gauss's law or electric fields: the electric flux through a closed surface is proportional to the total charge enclosed, Q.

In equation (1.8), I_d is called the *displacement current*, as it has units of Amps. However, I_d is not associated with an actual displacement of charge. It can be thought of as the current flowing through area **A** that would generate the same change in electric flux as **E** normal to that area. Thus, we can re-write Ampère's law by plugging equation (1.8) into equation (1.5).

$$\oint_C \mathbf{B} \cdot d\mathbf{l} = \mu_0(I + I_d). \tag{1.9}$$

In summary, Ampère's law states that a circulating magnetic field is generated by both an existing current, I, and displacement current I_d generated by an applied electric field.

1.3 Gauss's law for electric fields

Gauss's law for electric fields describes the flux through a closed surface S produced by a *static* electric field. The expression for Gauss's law is

$$\int_S \mathbf{E} \cdot \hat{\mathbf{n}} \; da = \frac{Q}{\varepsilon_0}, \tag{1.10}$$

which we already encountered in equation (1.8). In equation (1.10), $\hat{\mathbf{n}}$ is the unit vector normal to the surface, Q is the total charge *enclosed* by the surface, and ε_0, as usual, is the dielectric permittivity of free space, $\varepsilon_0 = 8 \cdot 85 \times 10^{-12}$ F/m.

In order to visualize the electric field, it is helpful to trace *field lines*. A field line is the trajectory that a positive charge would follow in a given electric field. For this visualization, we recall that the electrostatic force acting on a positive charge q exposed to the electric field E, is $\mathbf{F} = q\mathbf{E}$. Figure 1.8 illustrates several common examples of field lines associated with various electric fields. As expected, the field lines point away from a positive charge and toward a negative one (following the trajectory of hypothetical positive charge). Note that field lines do not cross, as this

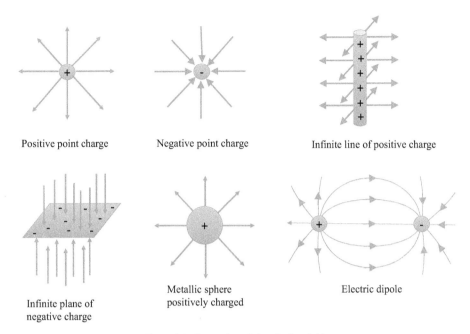

Positive point charge   Negative point charge   Infinite line of positive charge

Infinite plane of negative charge   Metallic sphere positively charged   Electric dipole

Figure 1.8. Examples of electric line fields.

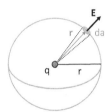

Figure 1.9. Example 1: using Gauss's law to calculating the electric field generated by a point charge at a distance **r**.

would imply, an electric field pointing in two different directions at a given point in space. Gauss' law allows us to easily calculate the electric field associated with a charge, especially when the charge is distributed across a domain with certain symmetry.

Example 1.1. Let us calculate the electric field generated by a point charge, q, at a certain distance r (figure 1.9).

Let us consider an imaginary spherical surface of radius r enclosing the charge. In order to calculate the electric flux (equation 1.10), we note that the element of area is

$$da = 4\,\pi r^2 dr \tag{1.11a}$$

And the normal at this area is along the position vector **r**,

$$\mathbf{n} = \frac{\mathbf{r}}{r}.$$

$$= \hat{\mathbf{r}}$$

(1.11*b*)

Thus, the electric flux is

$$\phi_E = \int_s \mathbf{E} \cdot \hat{\mathbf{n}} \ da$$

$$= 4\pi r^2 \mathbf{E} \cdot \hat{\mathbf{r}}.$$

(1.12*a*)

Using Gauss' law, we also know

$$\phi_E = \frac{q}{\varepsilon_0}.$$

(1.12*b*)

Multiplying equation (1.12b) by **r** and combining it with (1.12a), we finally obtain

$$\mathbf{E} = \frac{q}{4\pi\varepsilon_0 r^2}\hat{\mathbf{r}}.$$

(1.13)

It is left as an exercise to apply Gauss's law to the other geometries shown in figure 1.8 (see problem set, section 1.5).

Next, we discuss the analogous Gauss's law for magnetic fields.

1.4 Gauss's law for magnetic fields

While Gauss's law for magnetic fields is similar in form with the one for electric fields (section 1.3), the fundamental difference is that there is no magnetic equivalent for *charge*. In other words, while positive and negative electric charges can be detected separately, so far, the magnetic counterpart has not been found in nature. The magnetic poles ('north', 'south') always came in pairs. As a result, the right-hand side of Gauss's law vanishes, namely

$$\int_s \mathbf{B} \cdot \hat{\mathbf{n}} \ da = 0.$$

(1.14)

Equation (1.14) states that the magnetic flux passing through a closed surface is always zero. The analog to the electric force is called the magnetic or Lorentz force, defined as

$$\mathbf{F_B} = q\mathbf{v} \times \mathbf{B}$$

(1.15)

Equation (1.15). establishes that the force acting on a charge q moving at velocity **v** under the influence of magnetic field **B** is perpendicular to the plane (**v**, B), as illustrated in figure 1.10. Recall that, by contrast, the electric force is either parallel or antiparallel to the electric field, depending on the charge sign ($\mathbf{F} = q\mathbf{E}$).

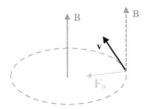

Figure 1.10. Magnetic force is perpendicular to the plane (**v**, **B**).

If θ is the angle between **v** and **B**, $\theta = \sphericalangle(\mathbf{v}, \ \mathbf{B})$, the magnitude of the magnetic field can be expressed as

$$B = \frac{F_B}{qv \ \sin \theta}, \tag{1.16}$$

where $B = |\mathbf{B}|$, $F_B = |\mathbf{F}_B|$, and $v = |\mathbf{v}|$.

We can see that the unit of B is $[B] = $ N/(c m s^{-1}) or Tesla (T).

We can easily find that the work performed by the magnetic force is always zero. Thus, using equation (1.15), we can calculate the elementary work corresponding to an infinitesimal displacement $d\mathbf{l}$, due to velocity $\mathbf{v} = d\mathbf{l}/dt$, as

$$\begin{aligned} dW_B &= \mathbf{F}_B \cdot d\mathbf{l} \\ &= q\left(\frac{d\mathbf{l}}{dt} \times \mathbf{B}\right) \cdot d\mathbf{l} \\ &= 0. \end{aligned} \tag{1.17}$$

In equation (1.17), we noted that the term in parenthesis is perpendicular to $d\mathbf{l}$. Unlike the electrostatic force, which is generated by static charges, the Lorentz force is created by charges in motion ($v \neq 0$). This is consistent with Ampère's law stating that the magnetic fields are generated by currents rather than charges (see equation (1.5)).

Due to the absence of magnetic monopoles, the magnetic field lines always form closed loops: they cannot converge to or emerge from a point (charge). Similar to the electric field lines, the magnetic ones also do not cross. Figure 1.11 shows examples of various magnetic field sources and their field lines.

1.5 Problems

1. Consider a time-varying, y-dependent magnetic field oriented along the x-axis, described by

$$\mathbf{B}(x, t) = B_0 y \ t \ \hat{\mathbf{x}}$$

Calculate the emf induced in a square loop of side a, placed at the plane $z = 0$ (figure 1.12). Show the direction of the induced current.

2. Redo Problem 1 for a magnetic field of the form

$$\mathbf{B}(x, t) = B_0 \cos(\Omega t)\hat{x}$$

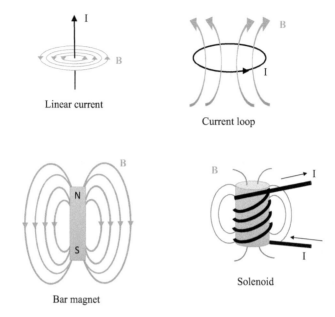

Figure 1.11. Examples of magnetic field lines.

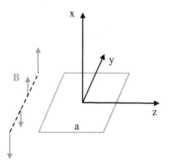

Figure 1.12. Problem 1.

3. Redo Problems 1–2 for a circular loop of radius a (figure 1.13).
4. A circular loop is rotating with the angular frequency Ω in a constant magnetic field $\mathbf{B}\|\mathbf{z}$, as shown in figure 1.14. The normal vector of the loop rotates in the x–z plane.

The magnetic field is 10^{-4} T, $\Omega = 2\,\pi\mathrm{rad}\;\mathrm{s}^{-1}$, the radius of the loop is $a = 1$ m, and the resistance of the of loop is $R = 100\,\Omega$.
   (a) Calculate the emf induced in the loop.
   (b) Calculate the induced current at $t = 125$ ms.
   5. Redo problem 4 for a square loop of $a = 1$ m.
   6. A circular loop is placed normal to a constant magnetic field B and its radius changes according to $r = r_0(1 + 0.5\cos \Omega t)$. Calculate the induced emf.

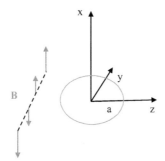

Figure 1.13. Problem 3.

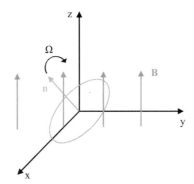

Figure 1.14. Problem 4.

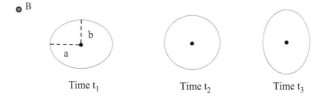

Figure 1.15. Problem 8.

7. Redo problem 6 for a square loop of side changing in time according to $r = r_0(1 + 0.5 \cos \Omega t)$.

8. An elliptical loop is placed normal to a constant magnetic field B. The loop is vibrating such that its major semi-axis changes in time as $a = a_0(1 + 05 \cos \Omega t)$. The length of the loop is constant in time and has a value of $L \simeq \pi(a + b)$, with a and b the two semi-axes (figure 1.15).

(a) Calculate the time-dependent minor semi-axis b.
(b) Calculate the induced emf.
(c) At $t = 0$, the loop starts rotating around a direction parallel to **B**. Calculate the induced emf.

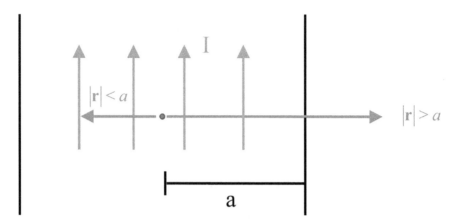

Figure 1.16. Problem 9.

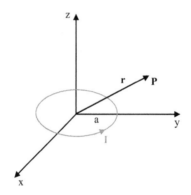

Figure 1.17. Problem 10.

(d) At $t = 0$, the loop starts rotating around a direction perpendicular to **B**. Calculate the induced emf for each possible direction of rotation.

9. Consider a linear wire of radius a carrying a constant current I (figure 1.16).
   (a) Calculate the induced magnetic field at a distance $r > a$.
   (b) Calculate the induced magnetic field at a distance $r < a$.

10. A circular loop of radius a carries a current I, as shown in figure 1.17. Calculate the induced magnetic field at a point P of coordinate **r** (figure 1.18).

11. A solenoid width N turns and radius a carries a current I (figure 1.7). Calculate the induced magnetic field.

12. Calculate the magnetic field generated at point P by an infinitesimal segment of a wire of arbitrary shape (figure 1.19).

13. Consider a linear wire carrying a current I. Calculate the magnetic flux through a rectangular surface positioned at $z = z_0$, of width Δz and length x_0, as in the geometry of figure 1.20.

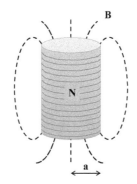

Figure 1.18. Problem 11.

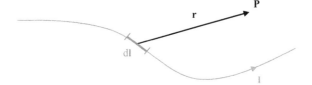

Figure 1.19. Problem 12.

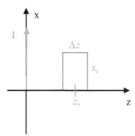

Figure 1.20. Problem 13.

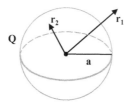

Figure 1.21. Problem 14.

14. Calculate the electric field generated by a conducting sphere of charge Q at a distance r, both inside and outside the sphere (figure 1.21).

15. Redo problem 14 for a dielectric sphere with the charge Q homogenously distributed.

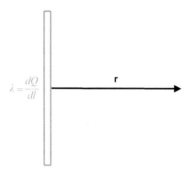

Figure 1.22. Problem 1.16.

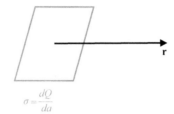

Figure 1.23. Problem 17.

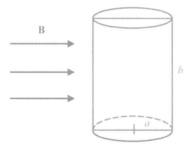

Figure 1.24. Problem 18.

16. Calculate the electric field at a distance **r** generated by a line charge of linear charge density λ (figure 1.22), when:

(a) the line is infinite in length;
(b) the line has a length L.

17. Calculate the electric field generated by an infinite plane, of surface charge density σ (figure 1.23), when:

(a) the plane is infinite;
(b) the plane is a square of side b.

18. Prove that the total magnetic flux through the cylindrical surface in figure 1.24 is zero.

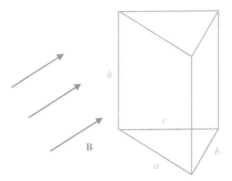

Figure 1.25. Problem 19.

19. Prove that the total magnetic flux through the prism in figure 1.25 is zero. Consider $\mathbf{B} = (B_x, B_y, B_z)$ along an arbitrary direction.

References and Further reading

[1] Faraday M 1832 *Experimental Researches in Electricity* (London: R. Taylor)
[2] Maxwell J C 1873 *A Treatise on Electricity and Magnetism* (Oxford: Clarendon)
[3] Fleisch D A 2008 *A Student's Guide to Maxwell's Equations* (Cambridge: Cambridge University Press)
[4] Kong J A 2008 *Electromagnetic Wave Theory* (Cambridge, MA: EMW Publishing)

IOP Publishing

Principles of Biophotonics, Volume 3
Field propagation in linear, homogeneous, dispersionless, isotropic media
Gabriel Popescu

Chapter 2

Maxwell's equations in differential form

Throughout this book, we will use Maxwell's equations in differential form, which are entirely equivalent to their integral counterparts described in chapter 1. Importantly, in the future will write explicitly the arguments of all the variables in these equations, such that there is no confusion as to whether we describe the problem in space, time, or their conjugate frequency domains.

2.1 The four main equations

The first Maxwell equation, *Faraday's induction law*, establishes that a varying magnetic field produces a *rotating or solenoidal* electric field,

$$\nabla \times \mathbf{E}(\mathbf{r},\, t) = -\frac{d\mathbf{B}(\mathbf{r},\, t)}{dt}. \tag{2.1}$$

In equation (2.1), $\mathbf{E}(\mathbf{r},\, t)$ is the instantaneous *electric field* (in V m^{-1}) and $\mathbf{B}(\mathbf{r},\, t)$ the instantaneous *magnetic induction* (in Tesla). Note that integrating equation (2.1). over an area S and using the Stokes theorem, namely, $\int_S \nabla \times \mathbf{E}(\mathbf{r},\, t) \cdot \hat{\mathbf{n}}\, da = \oint_C \mathbf{E} \cdot d\mathbf{l}$, we obtain Faraday's equation in integral form (equation (1.1)).

Ampère's circuital law (with Maxwell's correction) states that a magnetic field can be generated both by existing currents (the original Ampère's law) and by varying electric fields (Maxwell's correction),

$$\nabla \times \mathbf{H}(\mathbf{r},\, t) = \frac{d\mathbf{D}(\mathbf{r},\, t)}{dt} + \mathbf{j}(\mathbf{r},\, t). \tag{2.2}$$

In equation (2.2), $\mathbf{H}(\mathbf{r},\, t)$ is the instantaneous *magnetic field intensity* (A m^{-1}), $\mathbf{D}(\mathbf{r},\, t)$ the instantaneous *electric displacement, electric induction,* or *flux density* (C m^{-2}), and $\mathbf{j}(\mathbf{r},\, t)$ the instantaneous *source* current density (A m^{-2}). Note that the electric field itself can induce 'conductance' currents in conductive materials. These *induced*

doi:10.1088/978-0-7503-1646-0ch2

currents are contained in the electric displacement, **D**. Equation (2.2) also yields the integral form via the Stokes theorem.

Gauss's laws for the electric and magnetic fields are:

$$\nabla \cdot \mathbf{D}(\mathbf{r},\, t) = \rho(\mathbf{r},\, t) \tag{2.3}$$

where $\rho(r,\, t)$ is the volume charge density (Coulomb m^{-3}), and

$$\nabla \cdot \mathbf{B}(\mathbf{r},\, t) = 0 \,, \tag{2.4}$$

which establishes the non-existence of a magnetic charge. Equations (2.3) and (2.4) yield their integral counterparts, (1.10) and (1.14), via Gauss's theorem of vector algebra, which states that $\int_V \nabla \cdot \mathbf{D}(\mathbf{r},\, t)dV = \int_S \mathbf{D} \cdot \hat{\mathbf{n}} \; da$. Equations (2.1) and (2.4) represent the four main equations of electromagnetic theory. Various combinations of these equations can be obtained.

Interestingly, by taking the divergence of Ampère's law (equation (2.2)) and recalling that the divergence of a curl vanishes identically, we obtain the *continuity equation*,

$$\nabla \cdot \mathbf{j}(\mathbf{r},\, t) + \frac{\partial \rho(\mathbf{r},\, t)}{\partial t} = 0 \,. \tag{2.5}$$

Equation (2.5) simply establishes the conservation of electric charge.

In order to solve the system of Maxwell equations, we need to specify the electric and magnetic properties of the medium of interest. Mathematically, we need to determine relationships between **E** and **D**, and between **H** and **B**. As described in the next section, these *constitutive relations* allow us to connect the field distributions with the material properties. In biological investigations, we always extract information about the specimen of interest (e.g., structure, dynamics) by solving such 'optical' inverse problems.

2.2 Constitutive relations

The electric and magnetic inductances depend strongly on the respective properties of the medium. For example, the same electric field can induce different charge displacements in different materials, both in magnitude, phase, and orientation. Thus, in addition to the four Maxwell equations introduced above (equations (2.1)–(2.4)}, particular *constitutive relations* apply. For *isotropic* media, they are

$$\begin{aligned}\mathbf{D} &= \varepsilon\mathbf{E} \\ &= \varepsilon_0\mathbf{E} + \mathbf{P},\end{aligned} \tag{2.6a}$$

where

$$\varepsilon = \varepsilon_0\varepsilon_r \tag{2.6b}$$

and

$$\begin{aligned}\mathbf{B} &= \mu\mathbf{H} \\ &= \mu_0\mathbf{E} + \mathbf{M},\end{aligned} \tag{2.6c}$$

where

$$\mu = \mu_0\mu_r \tag{2.6d}$$

In equations (2.6*a*)–(2.6*d*), ε is the *dielectric permittivity*, μ the *magnetic permeability*, while **P** and **M** are the induced *polarization* and *magnetization*, respectively. The *relative* dielectric permittivity, ε_r, and magnetic permeability, μ_r, are defined with respect to the vacuum values, $\varepsilon_0 = 10^{-9}/36\,\pi$ F m^{-1} and $\mu_0 = 4\pi 10^{-7}$ A m^{-1}, respectively. Generally, these material quantities, ε, μ, are *anisotropic* (direction-dependent, see volume 4), *dispersive* (i.e., *t*- and ω-dependent, see volume 5 [1]) *inhomogeneous* (*r*- and **k**-dependent, see volume 6 [2]), and *nonlinear* (i.e., dependent on the input fields **E** and **H**, see volume 7 [3]).

Next, we discuss Maxwell's equations in all possible representations of space, time, spatial frequency, and temporal frequency domain.

2.3 Maxwell's equations in other representations

2.3.1 Space–frequency representation (r, ω)

For situations that involve (temporally) *broadband fields*, solving for the *spatial* behavior of *each temporal frequency* is often more efficient than calculating in the time domain. In order to obtain Maxwell's equations in the space–frequency representation, we use the *differentiation property* of Fourier transforms (see chapter 4 in volume 1 [4] for Fourier transform properties), namely,

$$\frac{d}{dt}\mathbf{F}(t) \leftrightarrow -i\omega \mathbf{F}(\omega) \tag{2.7}$$

In equation (2.7), **F**(*t*) is an arbitrary time-dependent vector function, and **F**(ω) its Fourier transform. *Note that we will denote the Fourier transform of a function by the same symbol, with the understanding that, for example, F(t) and F(ω) are two distinct functions.*

Taking the Fourier transform of equations (2.1)–(2.4) and using the differentiation property in equation (2.6), we can rewrite Maxwell's equations in the (*r*, ω) representation,

$$\begin{aligned}
\nabla \times \mathbf{E}(\mathbf{r}, \omega) &= i\omega \mathbf{B}(\mathbf{r}, \omega) \\
\nabla \times \mathbf{H}(\mathbf{r}, \omega) &= -i\omega \mathbf{D} + \mathbf{j}(\mathbf{r}, \omega) \\
\nabla \cdot \mathbf{D}(\mathbf{r}, \omega) &= \rho(\mathbf{r}, \omega) \\
\nabla \cdot \mathbf{B}(\mathbf{r}, \omega) &= 0
\end{aligned} \tag{2.8}$$

Similarly, the constitutive relations take the form

$$\mathbf{D}(\mathbf{r}, \omega) = \varepsilon \mathbf{E}(\mathbf{r}, \omega), \tag{2.9a}$$

where $\varepsilon = \varepsilon_0 \varepsilon_r$ and **D**(**r**, ω) = ε_0**E**(**r**, ω) + **P**(**r**, ω), and

$$\mathbf{B}(\mathbf{r}, \omega) = \mu \mathbf{H}(\mathbf{r}, \omega), \tag{2.9b}$$

where $\mu = \mu_0 \mu_r$ and **B**(**r**, ω) = μ_0**H**(**r**, ω) + **M**(**r**, ω).

Note that here we assumed constant material properties, i.e., linear, isotropic, homogeneous, dispersionless media, meaning that ε and μ are constant. In the subsequent volumes, we will study separately situations where each of these conditions is *not* fulfilled.

2.3.2 Wavevector–time representation (k, *t*)

Often, we deal with optical fields characterized by broad *wavevector (angular) distribution or spectra*. These fields can be decomposed in wavevectors, **k,** of different directions. The natural representation of the fields in this case is in the **k**-vector space. Recall that a differentiation property analog to equation (2.7) holds for the ∇ operator, as follows (see volume 1)

$$\nabla \times \mathbf{F}(\mathbf{r}) \rightarrow i\mathbf{k} \times \mathbf{F}(\mathbf{k})$$
$$\nabla \cdot \mathbf{F}(\mathbf{r}) \rightarrow i\mathbf{k} \cdot \mathbf{F}(\mathbf{k})$$

(2.10)

Note that there is sign change between the time and space differentiation property, which is consistent with our choice of the monochromatic plane wave as $\exp[-i(\omega t - \mathbf{k} \cdot \mathbf{r})]$.

The (**k**, *t*) representation of Maxwell's equations is obtained by re-writing equations (2.1)–(2.4) as

$$i\mathbf{k} \times \mathbf{E}(\mathbf{k}, t) = -\frac{d\mathbf{B}(\mathbf{k}, t)}{dt} \qquad (2.11a)$$

$$i\mathbf{k} \times \mathbf{H}(\mathbf{k}, t) = \frac{d\mathbf{D}(\mathbf{k}, t)}{dt} + \mathbf{j}(\mathbf{k}, t) \qquad (2.11b)$$

$$\mathbf{k} \cdot \mathbf{D}(\mathbf{k}, t) = \rho(\mathbf{k}, t) \qquad (2.11c)$$

$$\mathbf{k} \cdot \mathbf{B}(\mathbf{k}, t) = 0 \qquad (2.11d)$$

2.3.3 Wavevector–frequency representation (k, *ω*)

The (**k**, *ω*) representation of Maxwell's equations is obtained via the temporal Fourier transformation of equations (2.11a)–(2.11d). For media of no free charge ($\rho = 0$) or currents ($j = 0$), these equations are

$$\mathbf{k} \times \mathbf{E}(\mathbf{k}, \omega) = \omega\mathbf{B}(\mathbf{k}, \omega) \qquad (2.12a)$$

$$\mathbf{k} \times \mathbf{H}(\mathbf{k}, \omega) = -\omega\mathbf{D}(\mathbf{k}, \omega) \qquad (2.12b)$$

$$\mathbf{k} \cdot \mathbf{D}(\mathbf{k}, \omega) = 0 \qquad (2.12c)$$

$$\mathbf{k} \cdot \mathbf{B}(\mathbf{k}, \omega) = 0 \qquad (2.12d)$$

Equations (2.12a–d) describe the propagation of frequency components ω and plane wave of wavevectors **k**. Right away, equations (2.12c–d) establish that $\mathbf{k}\perp\mathbf{B}$ and $\mathbf{k}\perp\mathbf{D}$. Typically, μ is a scalar for most optical materials, such that $\mathbf{B}\|\mathbf{H}$.

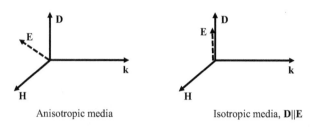

Anisotropic media   Isotropic media, **D||E**

Figure 2.1. Orientation of the **E, H, D**, and **k** vectors in anisotropic and isotropic media, as indicated.

However, for anisotropic materials, ε is a tensor, i.e., **D** is not necessarily parallel to **E**. From equation (2.12b) we see that **D⊥H**, such that, all together, we learned that

$$\begin{aligned} &\mathbf{k}\perp\mathbf{D} \text{ and} \\ &\mathbf{k}\perp\mathbf{H} \text{ and} \\ &\mathbf{H}\perp\mathbf{D}. \end{aligned} \qquad (2.13)$$

Equations (2.12a–d) indicate that **H, D**, and **k** are mutually orthogonal vectors (figure 2.1). For isotropic media, **D||E**, and **H, E**, and **k** are also mutually orthogonal.

2.4 Classification of optical materials

2.4.1 Anisotropic

For *anisotropic* media, the material properties depend on the direction of the electric and magnetic fields and, thus, are described by tensors, $\bar{\bar{\varepsilon}}$ and $\bar{\bar{\mu}}$. In this case, the constitutive relations are

$$\begin{aligned} \mathbf{D} &= \bar{\bar{\varepsilon}}\mathbf{E} \\ &= \begin{pmatrix} \varepsilon_{xx} & \varepsilon_{xy} & \varepsilon_{xz} \\ \varepsilon_{yx} & \varepsilon_{yy} & \varepsilon_{yz} \\ \varepsilon_{zx} & \varepsilon_{zy} & \varepsilon_{zz} \end{pmatrix} \begin{pmatrix} E_x \\ E_y \\ E_z \end{pmatrix} \end{aligned} \qquad (2.14a)$$

$$\begin{aligned} \mathbf{B} &= \bar{\bar{\mu}}\mathbf{H} \\ &= \begin{pmatrix} \mu_{xx} & \mu_{xy} & \mu_{xz} \\ \mu_{yx} & \mu_{yy} & \mu_{yz} \\ \mu_{zx} & \mu_{zy} & \mu_{zz} \end{pmatrix} \begin{pmatrix} H_x \\ H_y \\ H_z \end{pmatrix} \end{aligned} \qquad (2.14b)$$

The *induced polarization*, or the electric dipole moment per unit volume, can be further expressed in terms of a *dielectric susceptibility*, χ, which is also a tensor

$$\mathbf{P} = \varepsilon_0 \bar{\bar{\chi}}\, \mathbf{E} \qquad (2.15)$$

From equation (2.15), we see that

$$\bar{\bar{\chi}} = \bar{\bar{\varepsilon}}_r - 1. \qquad (2.16)$$

Equation (2.14a) shows that for electrically anisotropic media, the electric field, **E**, is no longer parallel with the induction, **D,** or, equivalently, the induced polarization is not parallel with **E**. Crystals are common anisotropic media, for which equation (2.14a) applies. As detailed further in volume 4, propagation in crystals is greatly simplified using the *principal coordinate system*, in which $\bar{\bar{\varepsilon}}$ becomes diagonal.

$$\bar{\bar{\varepsilon}} = \begin{pmatrix} \varepsilon_x & 0 & 0 \\ 0 & \varepsilon_y & 0 \\ 0 & 0 & \varepsilon_z \end{pmatrix}. \tag{2.17}$$

Depending on whether the values of ε_x, ε_y, and ε_z are all distinct, crystals fall into three groups, as follows.

Group 1: isotropic, $\varepsilon_x = \varepsilon_y = \varepsilon_z$.
In this type of crystal, the dielectric response is the same along the three axes. These crystals belong to the cubic system. (figure 2.2).

Group 2: uniaxial, $\varepsilon_x = \varepsilon_y \neq \varepsilon_z$.
In these crystals, the permittivity corresponding to the principal axes, say, x and y is the same, but different from the one along the third axis (z, in this case). These crystals belong to the *trigonal, tetragonal,* and *hexagonal* systems, as described in figure 2.2. Such crystals have a preferential direction along which a plane wave experiences the same refractive index irrespective of the orientation of the electric field, **E**. Thus, these crystals are called *uniaxial*.

Group 3: biaxial, $\varepsilon_x \neq \varepsilon_y \neq \varepsilon_z$.
This class of crystals belongs to the orthorombic, monoclinic, and triclinic systems (figure 2.2). Unlike Group 2, these crystals allow for two preferential directions along which a plane wave experiences the same refractive index, irrespective of the orientation of the electric field. Thus, they are called *biaxial*.

2.4.2 Dispersive

Virtually all optical materials exhibit *temporal dispersion*, also known as *color dispersion* or chromatic dispersion, i.e., are characterized by a dielectric constant that depends on (temporal) frequency, $\varepsilon(\omega)$. We will consider isotropic materials for now, whereby ε is a scalar, but the same phenomena exists in anisotropic materials, where each element of the tensor may depend on frequency, $\bar{\bar{\varepsilon}}(\omega)$.

In the frequency domain, the constitutive relation (2.6a) can be written as:

$$\mathbf{D}(\omega) = \varepsilon(\omega)\mathbf{E}(\omega). \tag{2.18}$$

Equation (2.18) indicates that, a given incident field Ecreates a frequency dependent electric displacement, **D**. Taking the Fourier transform inverse of equation (2.18), we obtain in the time domain the following expression

$$\mathbf{D}(t) = \int_{-\infty}^{\infty} \varepsilon(t')E(t - t')dt' \tag{2.19}$$
$$= \varepsilon(t) \circledS E(t),$$

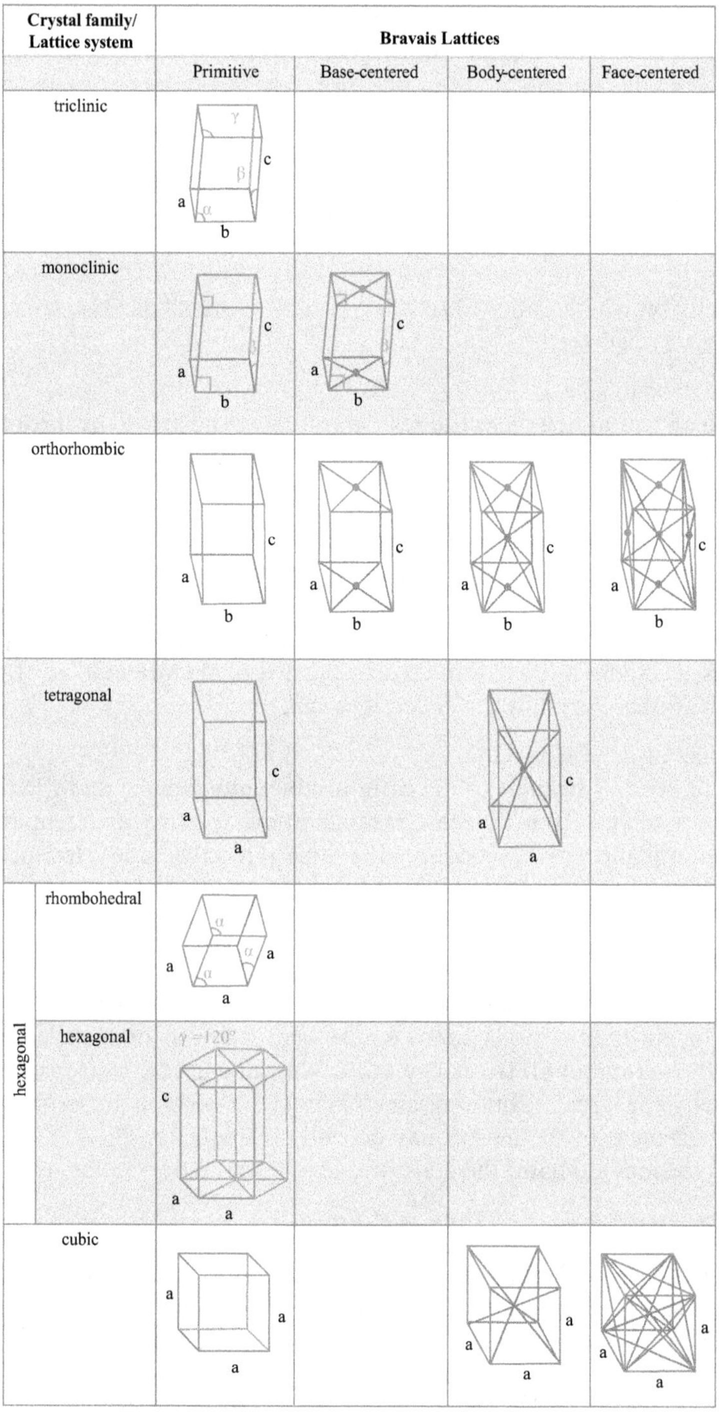

Figure 2.2. Classification of crystals by symmetry. There are 14 types of (Bravais) lattices, which describe the arrangement of the lattice points. Primitive is the repeating unit in a crystal.

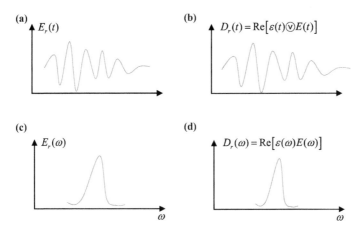

Figure 2.3. (a) Real part of an electric field. (b) Real part of the electric displacement in a medium of dielectric permittivity $\varepsilon(t)$. (c) Real part of the electric field in the frequency domain. (d) Real part of the electric displacement in the frequency domain as the result of a filtering operation with the material response $\varepsilon(\omega)$.

Where we used the same symbols in the time and frequency domain to express Fourier transform pairs, $f(t) \leftrightarrow f(\omega)$. In deriving equation (2.19), we used the convolution theorem (see volume 1 [4]), with $\odot$ denoting the convolution operation. The integral over t' in equation (2.19) indicates that the response of the material has a certain signature in time, like all linear systems. We note that if the material has an infinitely sharp response, of the form

$$\varepsilon(t) = \varepsilon_1 \delta(t) \ . \tag{2.20}$$

its frequency counterpart becomes constant,

$$\varepsilon(\omega) = \varepsilon_1 \ . \tag{2.21}$$

This result shows that the physical origin of temporal dispersion is the *non-instantaneous* response of the material to the incident electromagnetic field (figure 2.3(a) and (b)). Thus, dispersive materials act as frequency filters onto the incident light (figure 2.3(c) and (d)).

Temporal dispersion plays an important role in light pulse propagation in various materials, can be exploited in optical spectroscopy measurements, and may affect the quality of images obtained with broadband light. Propagation of light in dispersive media will be described in detail in volume 5 [1].

2.4.3 Inhomogeneous

Inhomogeneous (or heterogeneous) media are characterized by a dielectric permittivity that depends on the spatial coordinate, $\varepsilon(\mathbf{r})$. Such materials can be regarded as being *spatially dispersive*.

In the spatial domain, the electric displacement is

$$\mathbf{D}(\mathbf{r}) = \varepsilon(\mathbf{r})\mathbf{E}(\mathbf{r}), \tag{2.22}$$

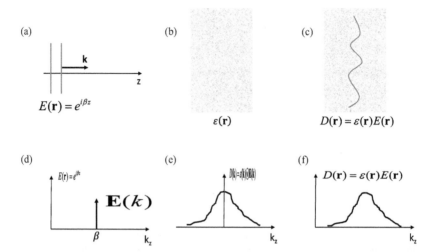

Figure 2.4. (a) Scalar plane wave propagating along z. (b) Inhomogeneous medium, characterized by $\varepsilon(\mathbf{r})$. (c) Electric displacement in the medium. (d) Fourier transform of the incident field. (e) Fourier transform of the dielectric susceptibility along k_z. (f) Electric displacement in the k_z domain.

where, for simplicity, we assumed ε as scalar (isotropic media). In the spatial frequency domain, the relationship becomes a convolution, namely,

$$\mathbf{D}(\mathbf{k}) = \varepsilon(\mathbf{k}) \otimes \mathbf{E}(\mathbf{k}). \tag{2.23}$$

Note the reversed relationships in (2.22)–(2.23) with respect to those in the equations (2.18) and (2.19), where the convolution occurs in the time domain. That is to say: while a dispersionless medium means $\varepsilon(\omega) = $ const., a homogeneous one implies $\varepsilon(\mathbf{r}) = $ const.. The latter is characterized by spatial frequency response in the form of a δ- function, $\varepsilon(\mathbf{k}) = \varepsilon_1 \delta(\mathbf{k})$.

Figure 2.4 illustrates how an incident plane wave interacts with an inhomogeneous medium. Spatial inhomogeneity gives rise to light scattering, which is the physical phenomenon fundamental to many imagining methods. Field propagation in inhomogeneous media will be treated in detail in volume 6 [2].

2.4.4 Nonlinear

A nonlinear optical material simply does not satisfy the conditions for a linear system (see volume 1, chapter 2). Specifically, the material response function depends on the input field. Thus, the electric displacement has the form

$$\mathbf{D} = \varepsilon(\mathbf{E})\mathbf{E}. \tag{2.24}$$

Using equation (2.15), we can express the induced polarization in terms of the dielectric susceptibility, χ, which is traditionally used to characterize nonlinear material properties,

$$\mathbf{D} = \varepsilon_0 \overline{\overline{\chi}}\, \mathbf{E}. \tag{2.25}$$

The susceptibility, $\overline{\overline{\chi}}$, is generally a tensor, which can be expanded in a series of the form

$$\overline{\overline{\chi}}(\mathbf{E}) = \overline{\overline{\chi^{(1)}}} + \overline{\overline{\chi^{(2)}}}\mathbf{E} + \overline{\overline{\chi^{(3)}}}\mathbf{EE} + \dots \tag{2.26}$$

In equation (2.26), we see that $\overline{\overline{\chi^{(1)}}}$ is nothing more than the linear, input-independent response of the material. The following terms denote the second-order nonlinearity ($\overline{\overline{\chi^{(2)}}}$), third-order ($\overline{\overline{\chi^{(3)}}}$), etc. The induced polarization can be written as

$$\mathbf{P} = \varepsilon_0\left[\overline{\overline{\chi^{(1)}}}\mathbf{E} + \overline{\overline{\chi^{(2)}}}\mathbf{EE} + \overline{\overline{\chi^{(3)}}}\mathbf{EEE} + \dots\right], \tag{2.27}$$

where the electric field products are computed as outer products and $\chi^{(2)}$ is a third-rank tensor, characterized by χ_{ijk}, i, j, $k = 1,2,3$, $\chi^{(3)}$ is a fourth-rank tensor, etc.

For example, one component of the induced polarization vector has the following expression for a second-order ($\chi^{(2)}$ or 'chi two') material (figure 2.4)

$$P_i^{(2)} = \varepsilon_0 \sum_{j,k}^{1,2,3} \chi_{ijk}^{(2)} E_j E_k. \tag{2.28}$$

Optical nonlinearities are generally small. For example, although predicted theoretically in the 1930s, experimental nonlinear effects were demonstrated only after the invention of lasers, which allowed for high irradiance levels to be delivered into the material (figure 2.5).

Today, there are many biophotonics tools based on the nonlinear interaction between the light and the specimen of interest. Through harmonic generation and other nonlinear effects (e.g., optical parametric generation, The principles of nonlinear optics will be discussed in more detail in volume 7 [3].

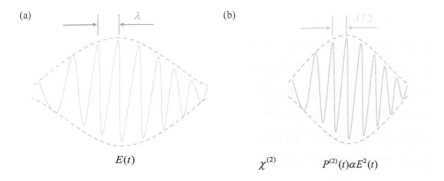

Figure 2.5. (a) Electric field in vacuum. (b) Induced polarization in a second-order material oscillates at double the freqeuncy.

2.5 Boundary conditions

Studying the behavior of the electric and magnetic fields at the boundary between two media allows us to understand basic phenomena, such as reflection and transmission at interfaces. In order to arrive at these *boundary conditions* for **E** and **H**, let us consider the interface between two media as the plane $z = 0$ (figure 2.6). A 'pill-box' region of infinitesimal thickness, $\Delta z \rightarrow 0$, is traversed by the plane $z = 0$. We recall Maxwell's second equation (Ampère's law) and seek to find the boundary conditions for the magnetic field,

$$\bar{\nabla} \times \mathbf{H}(\mathbf{r}, t) = \frac{d\mathbf{D}(\mathbf{r}, t)}{dt} + \mathbf{j} \ . \tag{2.29}$$

Because the size of the box in $x - y$ is much larger than in z, we can ignore partial derivatives along x and y and only keep those along z. Thus, $\bar{\nabla} \times \mathbf{H}$ can simplified as:

$$\bar{\nabla} \times \mathbf{H} = \left(0 \cdot \hat{x}, \ 0 \cdot \hat{y}, \ \frac{\partial}{\partial z}\hat{z}\right) \times \mathbf{H}$$

$$= \frac{\partial}{\partial z}(\hat{z} \times \mathbf{H})$$

$$= \lim_{\Delta z \to 0} \frac{\hat{z} \times \left[\mathbf{H}\left(x, y, \frac{\Delta z}{2}\right) - \mathbf{H}\left(x, y, -\frac{\Delta z}{2}\right)\right]}{\Delta z} \tag{2.30}$$

$$= \lim_{\Delta z}\left[\frac{\hat{z} \times (\mathbf{H_2} - \mathbf{H_1})}{\Delta z}\right].$$

In equation (2.30), $\mathbf{H_2} = \mathbf{H_2}(x, y, \Delta z/2)$ is the magnetic field in medium 2 and, $\mathbf{H_1} = \mathbf{H_1}(x, y, -\Delta z/2)$ in region 1. Eliminating $\nabla \times \mathbf{H}$ between equations (2.29) and (2.30), we obtain

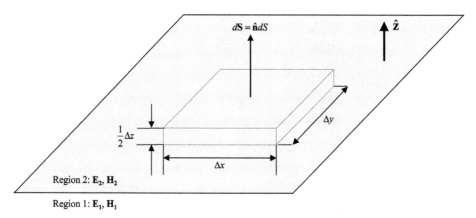

Figure 2.6. The basic phenomena of reflection and transmission at interfaces is better explained using a 'pill-box' region of infinitesimal thickness for studying the behavior of the electric and magnetic fields at the boundary between two media.

$$\hat{z} \times [\mathbf{H}_2 - \mathbf{H}_1] = \lim_{\Delta z \to 0} \Delta z \left[\frac{\partial \mathbf{D}}{\partial t} + \mathbf{j} \right] \tag{2.31}$$

In equation (2.31), we can assume $\delta \mathbf{D}/\delta t$ as finite, such that $\Delta z \delta D/\delta t \to 0$ when $\Delta z \to 0$. However, the current density can become infinite at the surface of perfect conductors. Thus, the last term on the RHS of equation (2.31) does not necessarily vanish and expresses the *surface* current density,

$$\mathbf{j}_s = \lim_{\Delta z \to 0} \mathbf{j} \Delta z. \tag{2.32}$$

Finally, we note that the unit vector $\hat{z}$ is the same as the normal unit vector at the boundary, $\hat{n}$. Thus, combining equations (2.31) and (2.32), we obtain the general boundary condition for the magnetic field, namely,

$$\hat{n} \times (\mathbf{H}_2 - \mathbf{H}_1) = \mathbf{j}_s. \tag{2.33}$$

Equation (2.33) expresses that the discontinuity in the *tangential* components of the magnetic field equal the surface current density.

Following the same procedure but starting with Faraday's equation, $\bar{\nabla} \times \mathbf{E} = -\partial \mathbf{B}/\partial t$, we readily obtain a similar boundary condition for the electric field,

$$\hat{n} \times (\mathbf{E}_2 - \mathbf{E}_1) = 0. \tag{2.34}$$

Equation (2.34) indicates that the tangential components of the electric field are continuous.

The Gauss equations for the electric and magnetic fields provide the starting point for deriving the behavior at the boundary $\mathbf{D}$ and $\mathbf{B}$ (figure 2.7). Consider that the medium in the pill-box of figure 2.6 contains free charges of density ρ Gauss's law for electric displacement reads

$$\bar{\nabla} \cdot \mathbf{D} = \rho. \tag{2.35}$$

Neglecting the in-plane partial derivatives, $\partial D_x/\partial x \simeq \partial D_y/\partial y = 0$, the divergence term can be simplified as earlier,

$$\bar{\nabla} \cdot \mathbf{D} = \lim_{\Delta z \to 0} \left[\frac{D_z(x, y, \Delta z/2) - D_z(x, y - \Delta z/2)}{D_z} \right]$$
$$= \lim_{\Delta z \to 0} \frac{1}{\Delta z} [\mathbf{z} \cdot (\mathbf{D}_2 - \mathbf{D}_1)] \ . \tag{2.36}$$

Eliminating $\bar{\nabla} \cdot \mathbf{D}$ between equations (2.35) and (2.36), we obtain

$$\mathbf{z} \cdot (\mathbf{D}_2 - \mathbf{D}_1) = \lim_{\mathbf{D} \to 0} \Delta z \rho. \tag{2.37}$$

As in the case of the current density, the volume charge density, ρ, may become infinite on the surface of a perfect conductor. Thus, we can define a surface charge density, $\rho_s = \lim_{\Delta z \to 0} \rho \Delta z$. Thus, equation (2.37) can be written as

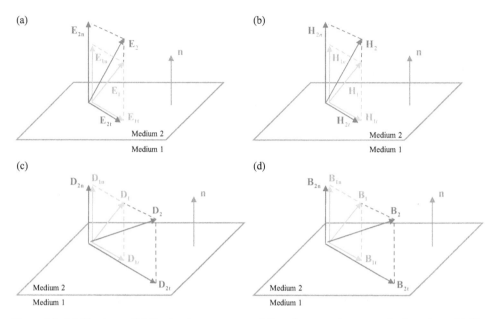

Figure 2.7. (a) The tangential (in-plane) components of **E** and are continuous at the boundary. (b) The tangential components of **H** are continuous (in the absence of surface currents). (c) The normal components of **D** and are continuous (in the absence of surface charge). (d) The normal components of **B** and are continuous.

$$\mathbf{n} \cdot (\mathbf{D}_2 - \mathbf{D}_1) = \rho_s \qquad (2.38)$$

where, again, we used $\hat{\mathbf{z}} = \hat{\mathbf{n}}$. Equation (2.38) indicates that the discontinuity of **D** along the normal direction amounts to the surface charge density, ρ_s.

The same reasoning can be easily applied to Gauss's law for magnetic fields, $\overline{\nabla} \cdot \mathbf{B} = 0$, to obtain

$$\hat{\mathbf{n}} \cdot (\mathbf{B}_2 - \mathbf{B}_1) = 0. \qquad (2.39)$$

Thus, we derived equations (2.33)–(2.34) that describe the tangential components of **E** and **H** at boundary, and equations (2.37)–(2.38) as the boundary conditions for the normal components of **D** and **B**.

In the absence of free surface charge and currents, we conclude that the tangential **E** and **H**, as well as the normal **D** and **B**, are all continuous across the boundary (figure 2.6).

$$\hat{\mathbf{n}} \times (\mathbf{E}_2 - \mathbf{E}_1) = 0 \qquad (2.40a)$$

$$\hat{\mathbf{n}} \times (\mathbf{H}_2 - \mathbf{H}_1) = 0 \qquad (2.40b)$$

$$\hat{\mathbf{n}} \cdot (\mathbf{D}_2 - \mathbf{D}_1) = 0 \qquad (2.40c)$$

$$\hat{\mathbf{n}} \cdot (\mathbf{B}_2 - \mathbf{B}_1) = 0 \qquad (2.40d)$$

2.6 Reflection and refraction at boundaries

2.6.1 Fresnel equations

Let us consider the problem of light propagation at the interface between two media (figure 2.8). We are interested to discover the rules that govern the **k**-vector direction change at the boundary between two media and the change in **E** and **H** upon reflection and transmission. In the following, we derive the expressions for the field reflection and transmission coefficients. In figure 2.8, the plane *x–z* is typically referred to as the *plane of incidence* (plane of the paper), i.e., the plane defined by the incident wavevector and normal ($\hat{\mathbf{n}}$) at the interface ($\hat{\mathbf{n}} \| \hat{\mathbf{z}}$). Recall the tangential components of the fields and normal components of inductions are conserved in the absence of surface currents and charges (equation (2.40)),

Since we are interested to study the **k**-vector direction change at the interface, the natural representation of Maxwell's equations is in the spatiotemporal frequency domain, **k**-ω. For $\rho = 0$ and $\mathbf{j} = 0$, Maxwell's equations in the **k**-ω representation yield, rearranging equations. (2.12a–b),

$$\mathbf{H}(\mathbf{k}, \omega) = \frac{1}{\omega\mu}\mathbf{k} \times \mathbf{E}(\mathbf{k}, \omega) \tag{2.41a}$$

$$\mathbf{E}(\mathbf{k}, \omega) = -\frac{1}{\omega\varepsilon}\mathbf{k} \times \mathbf{H}(\mathbf{k}, \omega) \tag{2.41b}$$

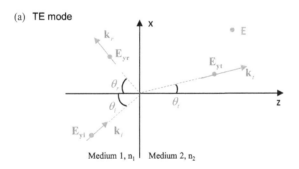

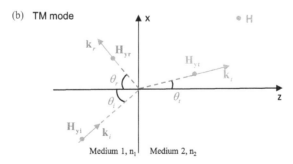

Figure 2.8. Geometry for the TE (a) and TM (b) modes. The plane of incidence is defined by the **k**-vectors and normal at the interface (*z*-axis in this case). Thus, TE and TM denote the electric and magnetic fields being transverse to the plane of incidence, i.e., parallel to the *y*-axis.

Expanding the cross products in equations (2.41a) and (2.41b) we find that the problem breaks down into two independent cases (modes): (a) *transverse electric (TE) mode*, when **E** is perpendicular to the plane of incidence (**E**||y, in figure 2.8(a) and (b) *transverse magnetic (TM) mode*, when **H**||y (figure 2.8(b)). We discuss these cases separately below. Note that the fields in medium 1 are the sum of the incident and reflected fields, while the ones in medium 2 consist of the transmitted fields only (figure 2.8).

i) TE mode (**E**||y)

If **E**||y, the boundary conditions for the tangent **E**-field and normal **H**-field (equation (2.40))

$$E_{yi} + E_{yr} = E_{yt} \qquad (2.42a)$$

$$H_{zi} + H_{zr} = H_{zt} \qquad (2.42b)$$

where the subscripts i, r, and t stand for *incident, reflected, and transmitted*. Using equation (2.41a) to express equation (2.42b) in terms of E_y components, we can rewrite the system of equations (equation (2.41a–b)) as

$$k_{xi}E_{yi} + k_{xr}E_{yr} = k_{xt}E_{yt} \qquad (2.43a)$$

$$E_{yi} + E_{yr} = E_{yt} \qquad (2.43b)$$

Since equations (2.43a) and (2.43b) must hold for any incident field E_{yi}, we obtain the following result

$$k_{xi} = k_{xr} = k_{xt}, \qquad (2.44)$$

which is known as the phase matching condition (figure 2.9). The result in equation (2.44) establishes a very basic result, known as *Snell's law*, which results immediately by noting that the x-component of the **k**-vector is $k \sin \theta$, namely,

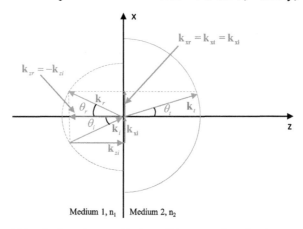

Figure 2.9. Phase matching (or **k**-vector matching) condition at an interface between two dielectrics: the tangential (within the interface) component of the **k**-vector is conserved for all three waves. The normal (perpendicular to the interface) component of the **k**-vector for the reflected field has opposite sign with respect to the normal component of the incident field.

$$n_1 \sin \theta_i = n_1 \sin \theta_r = n_2 \sin \theta_t. \qquad (2.45)$$

In equation (2.45) we used that the wavenumber, i.e., $k = |\mathbf{k}|$ in a medium of refractive index n relates to the wavenumber in vacuum as $k = n\beta_0$, $\beta_0 = \omega/c$.

Note that equation (2.45) implies $k_{zi} = -k_{zr}$, as the incident and reflection angles have opposite signs. In order to obtain the *field reflection coefficient*, we use the continuity of tangential $\mathbf{H}$ components,

$$H_{xi} + H_{xr} = H_{xt} \qquad (2.46)$$

which can be expressed in terms of $\mathbf{E}$ components (via equation (2.41b)),

$$k_{zi}E_{yi} + k_{zr}E_{yr} = k_{zt}E_{yt} \qquad (2.47)$$

Finally, combining equations (2.47) and (2.45a) to solve for the $\mathbf{E}$ field transmission and reflection coefficients, we obtain

$$r_{TE} = \frac{E_{yr}}{E_{yi}} = \frac{k_{zi} - k_{zt}}{k_{zi} + k_{zt}} \qquad (2.48a)$$

$$t_{TE} = \frac{E_{yt}}{E_{yi}} = \frac{2k_{zt}}{k_{zi} + k_{zt}} \qquad (2.48b)$$

Equations (2.48a) and (2.48b) are known as the *Fresnel Equations* for the TE mode (i.e., $\mathbf{E}\|\mathbf{z}$) and provide the field reflection and refraction coefficients when the refractive indices and incident angle are known. Note that, since coefficients r and t are ratios of fields and thus can have positive and negative values, in some situations, can even be complex.

We can express equation (2.48) in terms of the incident and refraction angles using the expression for the wavenumbers

$$k_i = n_1\beta_0 \qquad (2.49a)$$

$$k_r = n_1\beta_0 \qquad (2.49b)$$

$$k_t = n_2\beta_0 \qquad (2.49c)$$

Thus, expressing the projections of the wavevectors in polar coordinates, e.g., $k_x = k \sin \theta$, $k_z = k \cos \theta$. Thus, equation (2.48) can be re-written as

$$r_{TE} = \frac{n_1 \cos \theta_i - n_2 \cos \theta_t}{n_1 \cos \theta_i + n_2 \cos \theta_t} \qquad (2.50a)$$

$$t_{TE} = \frac{2n_1 \cos \theta_i}{n_1 \cos \theta_i + n_2 \cos \theta_t}$$
$$= 1 + r_{TE}. \qquad (2.50b)$$

The reflections and transmission coefficients in either form of (2.48) or (2.50) are known as the Fresnel equations. The conservation of energy dictates that the (power) reflectivity, R, and transmissivity, T, add up to unity,

$$R_{TE} + T_{TE} = 1, \qquad (2.51a)$$

$$R_{TE} = |\, n_{TE}\,|^2 \qquad (2.51b)$$

$$T_{TE} = |\, t_{TE}\,|^2 \qquad (2.51c)$$

It is left as exercise to show that the power conservation is indeed fulfilled.

*ii) TM mode (**H**∥y)*

Using the analog equations to the TE mode (equations (2.42a) and (2.42b)), i.e., the conservation of $\mathbf{H_y}$ components and normal $\mathbf{D}$ components, we find that k_x is conserved across the boundary for TM as well, as expected. This means that Snell's law (equation (2.45)) applies for both TE and TM modes. In order to obtain the field reflection and transmission coefficients, we use the conservation of both the tangent's fields components, E_x and H_y,

$$H_{yi} + H_{yr} = H_{yt}$$
$$\frac{k_{zi}}{n_1^2} H_{yi} + \frac{k_{zr}}{n_1^2} H_{yr} = \frac{k_{zt}}{n_2^2} H_{yt} \qquad (2.52)$$

The $\dfrac{1}{n_{1,2}^2}$ factor occurs due to the $1/\varepsilon$ factor in equation (2.41b) ($\varepsilon = n^2$). Thus, the $\mathbf{H}$ field reflection and transmission coefficients for the TM mode are

$$r_{TM} = \frac{H_{yr}}{H_{yi}} = \frac{\dfrac{k_{iz}}{n_1^2} - \dfrac{k_{tz}}{n_2^2}}{\dfrac{k_{iz}}{n_1^2} + \dfrac{k_{tz}}{n_2^2}} \qquad (2.53a)$$

$$t_{TM} = \frac{H_{yt}}{H_{yi}} = \frac{2k_{tz}}{\dfrac{k_{iz}}{n_1^2} + \dfrac{k_{tz}}{n_2^2}} \qquad (2.53b)$$

We expressed r_{TM} and t_{TM} in terms of H fields to emphasize the symmetry with respect to the TE case. Of course, the quantities can be further expressed in terms of E fields via $H = \dfrac{\mathbf{k} \times \mathbf{E}}{\omega\mu}$. Note that conservation of energy is satisfied in both cases,

$$|\, r_{TE}\,|^2 + |\, t_{TE}\,|^2 = 1$$
$$|\, r_{TM}\,|^2 + |\, t_{TM}\,|^2 = 1 \qquad (2.54)$$

Taken together, equations (2.48a) and (2.48b), and (2.53a) and (5.53b), referred to as the *Fresnel equations*, provide the reflected and transmitted fields for an arbitrary incident field. Because of the polarization dependence of the reflection and refraction coefficient, polarization properties of light can be modified via reflection and refraction. In the following we discuss two particular cases that follow from the Fresnel equations, where either the transmission or reflection coefficient vanishes.

2.6.2 Total internal reflection: critical angle

Setting the transmission coefficient to vanish, $t_{TE} = t_{TM} = 0$, yields the same condition of 'no transmission' in both TE and TM modes, i.e., $k_{zt} = 0$. Thus,

$$k_{zt} = \sqrt{k_t^2 - k_{xt}^2}$$
$$= \sqrt{n_2^2 \beta_0^2 - n_1^2 \beta_0^2 \sin^2 \theta_i} \qquad (2.55)$$
$$= 0$$

where we used the property that $k_x = $ constant at the interface (equation (2.44)), i.e. $k_{xt} = k_{xi}$. We see right away that equation (2.55) has a solution only if $n_2 < n_1$. Thus, the transmission vanishes for incidence angle θ_i larger than

$$\theta_c = \sin^{-1}\left(\frac{n_2}{n_1}\right) \qquad (2.56)$$

The angle θ_c is referred to as the *critical angle*, at which *total internal reflection* takes place. The total internal reflection can occur for both TE and TM polarizations, the only restriction being that $n_2 < n_1$, such that equation (2.56) allows for a real solution for θ_c. For angles of incidence that are larger than the critical angle, $\theta_i > \theta_c$, the TE electric field reflection coefficient becomes

$$r_{TE} = \frac{k_{zi} - i|k_{zt}|}{k_{zi} + i|k_{zt}|}$$
$$= \frac{e^{-i\phi_{TE}}}{e^{i\phi_{TE}}} = e^{-i2\phi_{TE}}, \qquad (2.57)$$

where $\phi_{TE} = \tan^{-1}\left(\frac{|k_{zt}|}{k_{zi}}\right)$. Therefore, for $\theta_i > \theta_c$, the reflection coefficient represents a pure phase shift, i.e., the power is 100% reflected, but the reflected field is shifted in phase by $2\phi TE$.

Similarly, for the TM mode we obtain

$$r_{TM} = e^{-i2\phi_{TM}}$$
$$\phi_{TM} = \tan^{-1}\left(\frac{|k_{zt}|}{k_{zi}}\right)\frac{n_1^2}{n_2^2} \qquad (2.58)$$

Since the ϕ_{TM} and ϕ_{TE} have different values, total internal reflection can be used to phase-shift the different components of the electric field by different amounts. This capability allows the polarization state of optical fields to be changed (for example, transform linear into circular polarization and vice-versa). Note that above the critical angle, the transmitted plane wave has the form

$$E_t = E_i e^{ik_{zt}z}$$
$$= E_i e^{-|k_{zt}|z} \qquad (2.59)$$

Equation (2.59) indicates that the field in medium 2 is decaying exponentially. Thus, the field is significantly attenuated over a distance on the order of $1/k_{zt}$, i.e., the field does not propagate, or is *evanescent*. Note that this phenomenon is very different from absorption, where the *power* rather than the *field* amplitude is attenuated upon propagation. In that case, the medium is characterized by a complex dielectric constant, $\varepsilon = \varepsilon' + i\varepsilon''$. Interestingly, this evanescent field can be converted into a propagating wave by bringing a dielectric very close to the interface, from medium 2. This 'tunneling' phenomenon has been exploited for high resolution imaging, as described later in volume 9 [7].

2.6.3 Total transmission: Brewster angle

Another particular case of Fresnel's equations is when the *reflection coefficient* vanishes.

For the *TE mode*, we have

$$
\begin{aligned}
r_{TE} &= 0 \\
k_{zi} &= k_{zt} \\
n_1 &= n_2.
\end{aligned}
\tag{2.60}
$$

Thus, for TE polarization, the only way to obtain 100% transmission through an interface is when there is no refractive index contrast between the two media, or when there is no interface at all, which is the *trivial solution*.

However, for the *TM mode* there is a nontrivial solution for obtaining a vanishing reflection coefficient,

$$
\begin{aligned}
r_{TM} &= 0 \\
k_{iz} &= k_{tz} \\
n_2 \cos \theta_i &= n_1 \cos \theta_t
\end{aligned}
\tag{2.61}
$$

The condition is satisfied simultaneously with Snell's law, such that we have

$$
\begin{aligned}
n_1 \sin \theta_i &= n_2 \sin \theta_t \\
n_2 \cos \theta_i &= n_1 \cos \theta_t
\end{aligned}
\tag{2.62}
$$

Multiplying the two equations in equation (2.62) side by side, we use $\sin(2\theta) = 2 \sin(\theta)\cos(\theta)$ and obtain

$$
\begin{aligned}
\sin 2\theta_i &= \sin 2\theta_t \\
\theta_i + \theta_t &= \frac{\pi}{2}
\end{aligned}
\tag{2.63}
$$

Physically, the absence of reflection at the Brewster angle for the TM polarization can be understood as the result of the induced polarization not radiating along its direction (figure 2.10(a)). Thus, the induced polarization $\mathbf{P}$ in medium 2 is parallel to $\mathbf{E_t}$. When $\theta_i + \theta_t = \frac{\pi}{2}$, the reflected and transmitted k-vectors would have to be

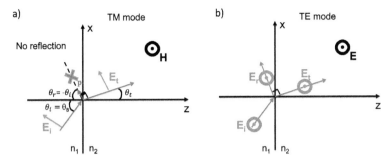

Figure 2.10. (a) Brewster's condition for TM polarization: when the reflected and transmitted directions are octagonal, the reflected field vanishes. p denotes the dipole induced by the incident field. The reflected field would propagate along p which is prohibited by electrodynamics (see chapter 3 for a fuller description of dipol radiation). (b) There is no Brewster condition for the TE polarization, as the induced dipole is perpendicular to the reflected field propagation direction, independent of the incidence angle.

perpendicular, and the reflected light direction would be parallel with **P**. This situation is prohibited by Maxwell's equations, resulting in the complete absence of the reflected field. As shown in figure 2.10(a), the relationship between the incidence and transmitted angle is

$$\theta_i + \theta_t = \frac{\pi}{2}. \tag{2.64}$$

The angle of incidence θ_B, referred to as the *Brewster angle*, for which $r_{TM} = 0$, is defined by combining equation (2.64) and Snell's law, namely,

$$n_1 \sin \theta_B = n_2 \sin\left(\frac{\pi}{2} - \theta_B\right), \tag{2.65}$$

or

$$n_1 \sin \theta_B = n_2 \cos(\theta_B), \tag{2.66}$$

which finally yields

$$\tan \theta_B = \frac{n_2}{n_1}. \tag{2.67}$$

Thus, unlike the total internal reflection, where the *transmission* can vanish for both polarizations, the *reflection* can only vanish for the TM mode. Figure 2.10(b) shows the situation for the TE polarization when the induced polarization are perpendicular to the reflected field for all incidence angles.

The Brewster angle is used in many applications, for example, to obtain polarized light from an incident *unpolarized* incident beam. If an unpolarized beam falls on a window at the Brewster angle, the reflected light is purely TE polarized, as the TM polarization is 100% transmitted. Windows at the Brewster angle are often used in laser cavities to obtain amplified polarized light.

2.7 Characteristic impedance

The characteristic *impedance*, η, of the medium is defined as the ratio of the transverse electric and magnetic field amplitudes,

$$\eta = \frac{E_x}{H_y}, \tag{2.68}$$

where we assume a plane wave propagating along z, $\mathbf{k} = (0,0, k)$, with the electric field along x, $\mathbf{E} = (E_x, 0,0)$, and magnetic field along y, $\mathbf{H} = (0, H_y, 0)$. Using the $(\mathbf{k}, \omega)$ representation to express the relationships between E_x and H_y (equations (2.11a) and (2.11b)), we obtain

$$kE_x(\mathbf{k}, \omega) = \omega\mu H_y(\mathbf{k}, \omega) \tag{2.69a}$$

$$- ikH_y(\mathbf{k}, \omega) = -i\omega\varepsilon E_x(\mathbf{k}, \omega) + \sigma E_x(\mathbf{k}, \omega) \tag{2.69b}$$

Taking the ratio of equations (2.69a) and (2.69b) eliminates k and yields immediately

$$\eta(\omega) = \sqrt{\frac{\mu}{\varepsilon + i\dfrac{\sigma}{\omega}}} \tag{2.70}$$

Interestingly, even for constant ε and μ, the impedance, η depends on frequency ω when the material has non-zero conductivity, i.e., $\sigma \neq 0$. The complex nature of η becomes physically more intuitive when we write it in *polar* form,

$$\eta(\omega) = \eta_0(\omega)e^{i\phi(\omega)}, \tag{2.71}$$

where the amplitude is given by

$$\eta_0(\omega) = \sqrt{\frac{\mu}{\varepsilon^2 + \left(\dfrac{\sigma}{\omega}\right)^2}} \tag{2.72}$$

and the phase by

$$\phi(\omega) = \tan^{-1}\left(-\frac{\sigma}{\varepsilon\omega}\right). \tag{2.73}$$

Thus, for a linearly polarized plane wave, the definition in equation (2.68) gives an intuitive relationship between the electric and magnetic fields,

$$\begin{aligned} E(\omega) &= \eta(\omega)H(\omega) \\ &= \eta_0(\omega)H(\omega)e^{i\phi(\omega)} \end{aligned} \tag{2.74}$$

Thus, the complex nature of the impedance indicates that, in addition to the amplitude relationship between $\mathbf{E}$ and $\mathbf{H}$, η controls the phase shift between the two fields. Note that for pure dielectrics $(\sigma = 0)$, $\eta = \sqrt{\mu/\varepsilon}$ and the frequency dependence is lost. In vacuum, the impedance has the value

$$\eta_0 = \sqrt{\frac{\mu_0}{\varepsilon_0}}$$
$$= 377 \ \Omega.$$

As $\omega \rightarrow 0$, we approach the 'electrostatics' case in which there is no propagation of electromagnetic field. Note that some textbooks show equation (2.70) with $-i\sigma/\omega$ in the denominator. This is because of the opposite sign adopted for monochromatic plane waves, $\exp(i\omega t - kz)$, instead of our $\exp(-i\omega t + kz)$.

2.8 Poynting theorem and energy conservation

In order to derive the equation that expresses the conservation of energy during field propagation, we start by dot-multiplying Faraday's law by $\mathbf{H}$ and Ampère's law by $\mathbf{E}$,

$$\mathbf{H}(\mathbf{r}, t) \cdot \nabla \times \mathbf{E}(\mathbf{r}, t) = -\mathbf{H}(\mathbf{r}, t) \cdot \frac{\partial B(\mathbf{r}, t)}{\partial t} \tag{2.75a}$$

$$\mathbf{E}(\mathbf{r}, t) \cdot \nabla \times \mathbf{H}(\mathbf{r}, t) = \mathbf{E}(\mathbf{r}, t) \cdot \frac{\partial \mathbf{D}(\mathbf{r}, t)}{\partial t} + \mathbf{E}(\mathbf{r}, t) \cdot \mathbf{j} \tag{2.75a}$$

Subtracting equation (2.75b) from (2.75a), we obtain

$$\nabla \cdot [\mathbf{E}(\mathbf{r}, t) \times \mathbf{H}(\mathbf{r}, t)] + \mathbf{H}(\mathbf{r}, t) \cdot \frac{\partial \mathbf{B}}{\partial t} + \mathbf{E} \cdot \frac{\partial \mathbf{D}}{\partial t} = -\mathbf{E}(\mathbf{r}, t) \cdot \mathbf{j}, \tag{2.76}$$

where we used the identity $\nabla \cdot [\mathbf{E}(\mathbf{r}, t) \times \mathbf{H}(\mathbf{r}, t)] = \mathbf{H} \cdot [\nabla \times \mathbf{E}(\mathbf{r}, t)] - \mathbf{E}(\mathbf{r}, t) \cdot [\nabla \times \mathbf{E}(\mathbf{r}, t)]$. Equation (2.76) is known as the *Poynting theorem*, which describes energy conservation. For real fields, the Poynting vector is defined as

$$\mathbf{S} = \mathbf{E}(\mathbf{r}, t) \times \mathbf{H}(\mathbf{r}, t) \tag{2.77}$$

which represents the flow of power density, $[S] = \frac{W}{m^2}$. If complex analytic signals are used instead, the Poynting vector is defined as $S = \frac{1}{2} \text{Re}[\mathbf{E} \times \mathbf{H}]$. The terms $\mathbf{E} \cdot \frac{\partial \mathbf{D}}{\partial t}$ and $\mathbf{H} \cdot \frac{\partial \mathbf{B}}{\partial t}$ represent, respectively, the rate of change of the stored electric field and magnetic energy. The term $-\mathbf{E} \cdot \mathbf{j}$ represents the power due to the source currents.

Let us consider the *real-valued* plane wave propagating in a homogenous medium along z,

$$E = \hat{x} A \cos(\omega t - kz)$$
$$H = \hat{y} \frac{A}{\eta} \cos(\omega t - kz)' \tag{2.78}$$

where η is the impedance. The *time-averaged* Poynting vector power density is

$$\langle \mathbf{S} \rangle_t = \hat{\mathbf{z}} \frac{A^2}{2\eta} \qquad (2.79)$$

The electric and magnetic power density are defined as

$$W_e = \frac{1}{2} \mathbf{E} \cdot \mathbf{D}$$
$$= \frac{1}{2} \varepsilon \mid \mathbf{E} \mid^2 \qquad (2.80a)$$

$$W_m = \frac{1}{2} \mathbf{H} \cdot \mathbf{B}$$
$$= \frac{1}{2} \mu \mid \mathbf{H} \mid^2 \qquad (2.80b)$$

Assuming no currents, $j = 0$, we can express the conservation of energy, or *Poynting's theorem* (equation (2.68)), as

$$\nabla \cdot \mathbf{S} + \frac{\partial}{\partial t}(W_e + W_m) = 0 \qquad (2.81)$$

where $W_e + W_m$ is the total energy density, $[W_e + W_m] = \dfrac{J}{m^3}$. Equation (2.81) expresses the conservation of energy; note that it takes the form of a continuity equation whereby the divergence of a current is balanced by a rate of change in energy density.

2.9 Phase, group, and energy velocity

Let us consider the electric field associated with a light beam propagating along the $+z$ direction with a wavevector $\mathbf{k} = (0,0,k)$. As illustrated in figure 2.11, the temporal signal is characterized by a slow modulation (*envelope*) due to the

(a)

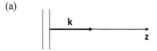

(b)

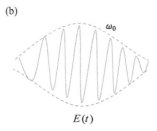

$E(t)$

Figure 2.11. Broad band plane wave: (a) propagation along z; (b) temporal distribution of the electric field (real part).

superposition of different frequencies, and a fast sinusoidal modulation (*carrier*), at the mean frequency ω_0,

$$E(z, t) = A(z, t)e^{-i(\omega_0 t - kz)} \tag{2.82}$$

Thus, the phase delay of the field is

$$\phi(z, t) = kz - \omega_0 t. \tag{2.83}$$

The *phase velocity* is associated with the advancement of *wave fronts*, which are defined by the distribution of points for which $\phi = $ constant. Differentiating equation (2.83), we obtain

$$\begin{aligned} d\phi(z, t) &= kdz - \omega_0 dt \\ &= 0 \end{aligned} \tag{2.84}$$

Thus, the phase velocity is

$$\begin{aligned} v_p &= \frac{dz}{dt} \\ &= \frac{\omega_0}{k} \\ &= \frac{c}{n}, \end{aligned} \tag{2.85}$$

where n is the refractive index of the medium, $k = n\omega/c$. Note that this simple expression holds for dispersionless media, i.e., n independent of ω.

For media with dispersion, the modulation (carrier) frequency is changing with time. Keeping in mind that $k = n(\omega)\omega/c$, we can expand the frequency $\omega(k)$ in Taylor series around the mean wavenumber k_0,

$$\omega(k) = \omega_0 + \frac{d\omega(k)}{dk}\bigg|_{k=k_0} (k - k_0) + \dots \tag{2.86}$$

The derivative on the right-hand side is the definition of the *group velocity*, v_g. Interestingly, expanding this derivative, we find that v_g relates to the phase velocity, following

$$\begin{aligned} \frac{d\omega(k)}{dk} &= \frac{d}{dk}\left(\frac{c}{n(\omega)}k\right) \\ &= \frac{c}{n(\omega)} - \frac{ck}{n(\omega)^2}\frac{dn(\omega)}{dk} \\ &= v_p\left(1 - \frac{\omega}{n}\frac{dn(\omega)}{d\omega}\right) \\ &= v_g(\omega) \end{aligned} \tag{2.87}$$

Equation (2.87) gives an expression for the *group velocity*, v_g, which is a function of frequency. The group velocity represents the velocity at which the envelope of the signal propagates. Note that the group and phase velocities are equal in non-dispersive media, when $\frac{dn}{d\omega} = 0$.

In the following, we prove that the *group velocity* is in fact the *energy velocity* of the field. By definition, the energy velocity is the ratio between the Poynting vector **S** and electromagnetic volume density U,

$$v_e = \frac{1}{U}\mathbf{S}$$
$$\mathbf{S} = \mathbf{E} \times \mathbf{H}$$
$$U = \frac{1}{2}\mathbf{E} \cdot \mathbf{D} + \frac{1}{2}\mathbf{H} \cdot \mathbf{B} \qquad (2.88)$$
$$= W_e + W_m$$

Further, we recall the source-free Faraday and Ampère's law in the k-ω representation (equation (2.12a–b)), namely, $\mathbf{k} \times \mathbf{E}(\mathbf{k}, \omega) = \omega\mathbf{B}(\mathbf{k}, \omega)$, $\mathbf{k} \times \mathbf{H}(\mathbf{k}, \omega) = -\omega\mathbf{D}(\mathbf{k}, \omega)$, which we differentiate on both sides to obtain

$$d\mathbf{k} \times \mathbf{E} + \mathbf{k} \times d\mathbf{E} = d\omega\mu\mathbf{H} + \omega\mu d\mathbf{H} \qquad (2.89a)$$

$$d\mathbf{k} \times \mathbf{H} + \mathbf{k} \times d\mathbf{E} = -d\omega\mu\mathbf{E} - \omega\mu d\mathbf{E} \qquad (2.89b)$$

Note that the meaning of $d\mathbf{k}$ and $d\omega$ is that of an infinitesimal spread in **k**-vectors (directions) and optical frequency, respectively. Let us dot-multiply equation (2.89a) by **H** and equation (2.89b) by **E**, such that we obtain

$$d\mathbf{k} \cdot (\mathbf{E} \times \mathbf{H}) + \mathbf{k} \cdot (\mathbf{H} \times d\mathbf{E}) = \mu d\omega\mathbf{H} \cdot \mathbf{H} + \omega\mu\mathbf{H} \cdot d\mathbf{H} \qquad (2.90a)$$

$$d\mathbf{k} \cdot (\mathbf{E} \times \mathbf{H}) + \mathbf{k} \cdot (d\mathbf{H} \times \mathbf{E}) = -d\omega(\mathbf{E} \cdot \varepsilon\mathbf{H}) - \omega(\mathbf{E} \cdot \varepsilon d\mathbf{E}) \qquad (2.90b)$$

Adding equations (2.90a) and (2.90b), we obtain

$$d\mathbf{k} \cdot (\mathbf{E} \times \mathbf{H}) = \frac{1}{2}d\omega[\mathbf{E} \cdot \varepsilon\mathbf{E} + \mu\mathbf{H} \cdot \mathbf{H}] \qquad (2.91)$$

In arriving at equation (2.91), we used that the products $d\mathbf{H} \cdot (\omega\mu\mathbf{H} - \mathbf{k} \times \mathbf{E})$ and $d\mathbf{H} \cdot (\varepsilon\mathbf{E}\omega + \mathbf{k} \times \mathbf{H})$ vanish, because $d\mathbf{H}{\perp}\mathbf{H}$, $d\mathbf{H}{\perp}(\mathbf{k} \times \mathbf{E})$, $d\mathbf{E}{\perp}\mathbf{E}$, and $d\mathbf{E}{\perp}(\mathbf{k} \times \mathbf{E})$.

Finally, using the definition in equation (2.80a) for the energy velocity, we obtain

$$v_e = \frac{\mathbf{E} \times \mathbf{H}}{(\mathbf{E} \cdot \varepsilon\mathbf{E} + \mu\mathbf{H} \cdot \mathbf{H})/2}$$
$$= \frac{d\omega}{d\mathbf{k}} \qquad (2.92)$$
$$= v_g$$

Equation (2.92) establishes the anticipated result that the electromagnetic energy flows at the group velocity. Note that the meaning of $\frac{\delta\omega}{\delta\mathbf{k}}$ is that of a gradient,

$\nabla_{\mathbf{k}} \omega = \left(\dfrac{\partial \omega}{k_x}, \dfrac{\partial \omega}{k_y}, \dfrac{\partial \omega}{k_y} \right)$. In *anisotropic* materials the group velocity can have different values along different directions and can even have different directions to the phase velocity.

It is apparent from equation (2.87) that v_g can exceed the speed of light in vacuum c, in special circumstances, of *anomalous dispersion*, when $\dfrac{dn}{d\omega} < 0$. However, this does not pose a conflict with the postulate of relativity theory, which states that the *signal velocity*, at which information can be transmitted via electromagnetic fields, is bounded by c. We will discuss dispersive materials in more detail in volume 6 [2].

2.10 The wave equation

Generally, eliminating **D**, **H**, and **B** among Maxwell's equations renders a vector equation in **E**, which will be referred to as the *(vector) wave equation*. However, in many biological applications the vector nature of the field, or light polarization, is not crucial. In this case, a *scalar wave equation* can be used. *Electromagnetic fields*, typically denoted by *E*, obey the vector wave equations and *scalar fields*, commonly denoted by *U*, obey the scalar wave equation. The two equations will be discussed separately next.

2.10.1 Vector wave equation

We recall Maxwell's equations (section 2.1), which, for media of constant dielectric permittivity, ε, non-magnetic materials, $\mu = \mu_0$, and no free charges, $\rho = 0$, have the form

$$\nabla \times \mathbf{E}(\mathbf{r}, t) = -\mu_0 \frac{\partial}{\partial t} \mathbf{H}(\mathbf{r}, t) \tag{2.93a}$$

$$\nabla \times \mathbf{H}(\mathbf{r}, t) = \varepsilon \frac{\partial}{\partial t} \mathbf{E}(\mathbf{r}, t) + \mathbf{j}(\mathbf{r}, t) \tag{2.93b}$$

$$\nabla \cdot \mathbf{H}(\mathbf{r}, t) = 0 \tag{2.93c}$$

$$\nabla \cdot \mathbf{E}(\mathbf{r}, t) = 0 \tag{2.93d}$$

In order to eliminate **H** from equations (2.93a) and (2.93b), we apply the curl operator to equation (2.93a) and plug in the expression for $\nabla \times \mathbf{H}$ from (2.93b), which results in

$$\nabla \times \nabla \times \mathbf{E}(\mathbf{r}, t) = -\mu_0 \frac{\partial}{\partial t} \nabla \times \mathbf{H}(\mathbf{r}, t)$$
$$= -\mu_0 \varepsilon \frac{\partial^2}{\partial t^2} \mathbf{E}(\mathbf{r}, t) - \mu_0 \frac{\partial}{\partial t} \mathbf{j}(\mathbf{r}, t). \tag{2.94}$$

The dielectric constant is generally complex, $\varepsilon \in \mathbb{C}$, indicating that the medium can interact with the light both *elastically*, via $\varepsilon' = \mathrm{Re}\,(\varepsilon)$, and *inelastically*, via $\varepsilon'' = \mathrm{Im}(\varepsilon)$,

$$\varepsilon = \varepsilon' + i\varepsilon''. \tag{2.95}$$

Since $i = e^{i\pi/2}$, we see that ε' defines the 'in-phase', *refractive* response of the medium, while ε'' describes the 'quadrature,' *loss* response,

$$\begin{aligned} \mathbf{D} &= \varepsilon\mathbf{E} \\ &= \varepsilon'\mathbf{E} + \varepsilon''\mathbf{E} \cdot e^{i\pi/2}. \end{aligned} \tag{2.96}$$

Sometimes ε is expressed in polar coordinates,

$$\begin{aligned} \varepsilon &= \sqrt{\varepsilon'^2 + \varepsilon''^2} \cdot e^{i\delta} \\ \delta &= \arctan\frac{\varepsilon''}{\varepsilon'}, \end{aligned} \tag{2.97}$$

where $\tan\delta$ is referred to as the 'loss tangent'.

It is physically revealing to introduce explicitly the real and imaginary parts of ε (equation (2.95)) in the wave equation (equation (2.94)),

$$\nabla \times \nabla \times \mathbf{E}(\mathbf{r},\, t) + \mu_0\varepsilon'\frac{\partial^2}{\partial t^2}\mathbf{E}(\mathbf{r},\, t) = -i\mu_0\varepsilon''\frac{\partial^2}{\partial t^2}\mathbf{E}(\mathbf{r},\, t) - \mu_0\frac{\partial}{\partial t}\mathbf{j}(\mathbf{r},\, t). \tag{2.98}$$

The first term on the RHS represents the loss due to *conduction currents*, or induced currents, also known in various contexts as eddy currents or Foucault currents. The RHS can be expressed in terms of a total current, i.e., the sum of the conduction, j_c, and source current, j, namely,

$$\nabla \times \nabla \times \mathbf{E}(\mathbf{r},\, t) + \mu_0\varepsilon'\frac{\partial^2}{\partial t^2}\mathbf{E}(\mathbf{r},\, t) = -\mu_0\frac{\partial}{\partial t}\big[j_c(\mathbf{r},\, t) + j(\mathbf{r},\, t)\big]. \tag{2.99}$$

The conduction current, j_c, is related to the conductivity, σ, $j_c = \sigma\mathbf{E}$, thus we can express it in two ways

$$\begin{aligned} j_c(\mathbf{r},\, t) &= i\varepsilon''\frac{\partial}{\partial t}\mathbf{E}(\mathbf{r},\, t) \\ &= \sigma\mathbf{E}(\mathbf{r},\, t). \end{aligned} \tag{2.100}$$

Equation (2.100) allows us to relate the loss dielectric response, ε'', to the conductivity. Thus, if we take the Fourier transform of equation (2.100), and use $\frac{\partial}{\partial t} \to -i\omega$, we obtain

$$\varepsilon'' = \frac{\sigma}{\omega}. \tag{2.101}$$

We can conclude that the $\pi/2$ out of phase, or loss response of the medium relates to the rate of change of the electric field, $\frac{\partial\mathbf{E}}{\partial t}$, and is due to the conductivity of the medium. The ratio between ε' and ε'' defines whether the medium is a good *conductor* or *insulator*,

$$\begin{aligned} \varepsilon'' \gg \varepsilon' \;(\sigma \gg \omega\varepsilon'), &\text{ good conductor} \\ \varepsilon' \gg \varepsilon'' \;(\sigma \ll \omega\varepsilon'), &\text{ good insulator (dielectric).} \end{aligned} \tag{2.102}$$

Equation (2.102) represents the vector wave equation, which is used to solve for the field propagation in anisotropic media, as detailed later, in volume 4 [8].

2.10.2 Scalar wave equation

Using the vector identity (see volume 1, appendix B) $\nabla \times \nabla \times \mathbf{E} = \nabla\nabla(\mathbf{E}) - \nabla^2\mathbf{E}$, we can expand the first term in wave equation (equation (2.90)), where the last term on the right-hand side is the Laplacian of the field. The $\nabla\nabla$ operator is a *dyadic operator*, obtained by the outer product of the ∇-operator, and has the form $\left(\dfrac{\partial}{\partial x_i} \quad \dfrac{\partial}{\partial x_j} \right)_{i,\,j=1,2,3}$. Throughout this book, we will denote the dyadic operators as a double overbar, $\overline{\overline{\nabla\nabla}}$, such that

$$
\overline{\overline{\nabla\nabla}} = \begin{pmatrix} \dfrac{\partial^2}{\partial x^2} & \dfrac{\partial^2}{\partial x \partial y} & \dfrac{\partial^2}{\partial x \partial z} \\[2mm] \dfrac{\partial^2}{\partial y \partial x} & \dfrac{\partial^2}{\partial y^2} & \dfrac{\partial^2}{\partial y \partial z} \\[2mm] \dfrac{\partial^2}{\partial z \partial x} & \dfrac{\partial^2}{\partial z \partial y} & \dfrac{\partial^2}{\partial z^2} \end{pmatrix}, \tag{2.103}
$$

Often, we can assume that the electric field is 'smooth', or slowly varying, over distances of the order of the wavelength, such that we can approximate $\nabla \cdot \mathbf{E} \simeq 0$, which renders the vector wave equation in a simplified form

$$
\nabla^2\mathbf{E}(\mathbf{r},\,t) - \mu_0\varepsilon\frac{\partial^2}{\partial t^2}\mathbf{E}(\mathbf{r},\,t) = \mu_0\frac{\partial}{\partial t}\mathbf{j}(\mathbf{r},\,t). \tag{2.104}
$$

Note that the RHS of equation (2.104) is the *driving term*, or the source of the electromagnetic field. Whether we deal with a macroscopic antenna or a fluorescent molecule, an existing variable current generates an electromagnetic field.

In lossless media (insulators), with no source currents, the wave equation simplifies further to

$$
\nabla^2\mathbf{E}(\mathbf{r},\,t) - \frac{n^2}{c^2}\frac{\partial^2}{\partial t^2}\mathbf{E}(\mathbf{r},\,t) = 0, \tag{2.105}
$$

where $n = \sqrt{\varepsilon'}$ is the refractive index and c the speed of light in vacuum.

In equation (2.91), the Laplacian operating on vector $\mathbf{E}$ is

$$
\begin{aligned}
\nabla^2\mathbf{E} &= \left(\frac{\partial^2}{\partial x^2} + \frac{\partial^2}{\partial y^2} + \frac{\partial^2}{\partial z^2} \right)\mathbf{E} \\[2mm]
&= \sum_{n=x,y,z} \left(\frac{\partial^2}{\partial x^2} + \frac{\partial^2}{\partial y^2} + \frac{\partial^2}{\partial z^2} \right)E_n\hat{\mathbf{n}}
\end{aligned} \tag{2.106}
$$

If we consider just one component of the electric field, we obtain the scalar equation as

$$\nabla^2 U(\mathbf{r},\, t) - \frac{n^2}{c^2}\frac{\partial^2 U(\mathbf{r},\, t)}{\partial t^2} = 0. \tag{2.107}$$

Next, we discuss the vector and scalar wave equation in all possible representations of space, time, spatial frequency, and temporal frequency domain.

2.11 Wave equation in other representations

2.11.1 Space–frequency representation (r, ω)

Taking the Fourier transform with respect to time of both the vector wave equation (equation (2.99)) and scalar wave equation (equation (2.107)), we obtain

$$\nabla \times \nabla \times \mathbf{E}(\mathbf{r},\, \omega) - \omega^2 \mu_0 \varepsilon' \mathbf{E}(\mathbf{r},\, \omega) = i\omega\mu_0 \big[j_c(\mathbf{r},\, \omega) + j(\mathbf{r},\, \omega) \big] \tag{2.108a}$$

$$\nabla^2 U(\mathbf{r},\, \omega) + n^2 \beta_0^2 U(\mathbf{r},\, \omega) = 0, \tag{2.108b}$$

where $\beta_0 = \omega/c$. Equation (2.108b) is generally referred to as the *Helmholtz equation*. The (r, ω) representation is particularly useful when dealing with non-monochromatic (temporally broadband) fields, as described in detail in volume 5 [1].

2.11.2 Wavevector–time representation (k, t)

Taking the Fourier transform with respect to *r* of both the vector wave equation (equation (2.99)) and scalar wave equation (equation (2.106)), we obtain

$$-\mathbf{k} \times \mathbf{k} \times \mathbf{E}(\mathbf{k},\, t) + \mu_0 \varepsilon' \frac{\partial^2}{\partial t^2}\mathbf{E}(\mathbf{k},\, t) = -\mu_0 \frac{\partial}{\partial t}\big[j_c(\mathbf{k},\, t) + j(\mathbf{k},\, t) \big] \tag{2.109a}$$

$$k^2 U(\mathbf{k},\, t) + \frac{n^2}{c^2}\frac{\partial^2 U(\mathbf{k},\, t)}{\partial t^2} = 0, \tag{2.109b}$$

where $k^2 = \mathbf{k} \cdot \mathbf{k} = |\mathbf{k}|^2$. This representation is useful when dealing with fields that are spatially broadband.

2.11.3 Wavevector–frequency representation (k, ω)

If we take the Fourier transform of the wave equations with respect to both **r** and t, we obtain the (k, ω) representation

$$\mathbf{k} \times \mathbf{k} \times \mathbf{E}(\mathbf{k},\, \omega) + \omega^2 \mu_0 \varepsilon' \mathbf{E}(\mathbf{k},\, \omega) = -i\omega\mu_0 \big[j_c(\mathbf{k},\, \omega) + j(\mathbf{k},\, \omega) \big] \tag{2.110a}$$

$$k^2 U(\mathbf{k},\, \omega) - n^2 \beta_0^2 U(\mathbf{k},\, \omega) = 0, \tag{2.110b}$$

Equation (2.110b) is satisfied for any field, thus, necessarily

$$k^2 = n^2 \frac{\omega^2}{c^2}. \tag{2.111}$$

Equation (2.111), known as the *dispersion relation*, relates the magnitude of the wavevector, $k = |\mathbf{k}|$, and the *wavenumber*, or *propagation constant*, $\beta_0 = n\frac{\omega}{c}$, which is a property of the material. Note that if the material is dispersive, $n(\omega)$, and inhomogeneous, $n(\mathbf{k})$, this dispersion relation, $k(\omega)$, becomes more complicated.

2.12 Problems

1. A magnetic field has following space-time dependence

$$\mathbf{B}(z, t) = B_0 \sin(-\omega t + kz)\hat{\mathbf{z}}.$$

   a) Calculate the curl of the induced electric field, $\nabla \times \mathbf{E}(\mathbf{r}, t)$.
   b) If $E_y = 0$, find E_x.
2. The induced electric field in a certain region is given by $\mathbf{E}(\mathbf{r}, t) = E_0(az\hat{\mathbf{x}} + bx\hat{\mathbf{y}} + cy\hat{\mathbf{z}})\cos \omega t$, with a, b, c constants. Calculate the rate of change of the magnetic field, $d\mathbf{B}(\mathbf{r}, t)/dt$.
3. A linear wire induces a magnetic field of the form (figure 2.12) $B = \frac{\mu_0 Ir}{2\pi a^2}\hat{\boldsymbol{\varphi}}$, where $\mathbf{r}$ is the coordinate of the observation point and a is the wire radius. Calculate the current density both inside and outside the wire.
4. A capacitor C with disk plates of radius a is charged by a voltage of potential difference V_0 through a wire of resistance R. The induced magnetic field has the form

$$\mathbf{B}(\mathbf{r}, t) = \frac{\mu_0 V_0}{2\pi R} e^{-\frac{t}{RC}} \left(\frac{r}{a^2}\right)\hat{\boldsymbol{\varphi}}$$

   Find the rate of change of the electric displacement, $\frac{d\mathbf{D}}{dt}$, between the capacitor plates.
5. The parallel wire separated by a distance a carries currents I and $10I$, in opposite directions (figure 2.13).

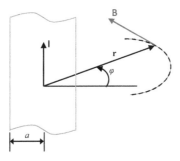

Figure 2.12. Problem 3.

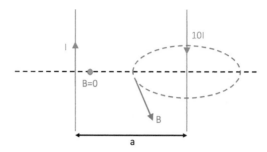

Figure 2.13. Problem 5.

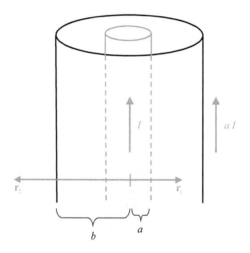

Figure 2.14. Problem 6.

   a) Find the induced magnetic field at a point of equal distance between
      the wires.
   b) Find the point of which the magnetic field vanishes.
6. The coaxial cable in figure 2.14 carries a current I on the inner conductor
   and aI on the outer conductor.
   a) Calculate the magnetic field at a distance r_1 between the two
      coordinates.
   b) Calculate the magnetic field at a distance r_2 outside the cable.
7. A current density induces a magnetic field

$$\mathbf{B}(\mathbf{r},\,t) = B_0 e^{ay}\cos(ax)\cos \omega t\hat{\mathbf{z}}.$$

Calculate the current density.
8. What is the current density that induces the magnetic field given by (use
   cylindrical coordinates)

$$\mathbf{B}(r) = B_0 e^{-ar}\cos y\hat{\mathbf{z}}.$$

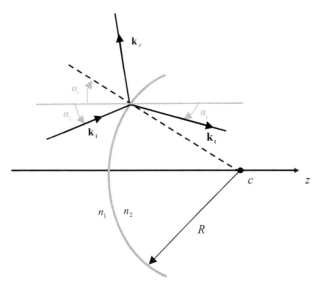

Figure 2.15. Problem 9.

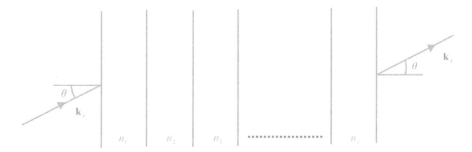

Figure 2.16. Problem 10.

9. A plane wave is incident on a spherical surface of radius R, as shown in figure 2.15. The surface separates two media of refractive index n_1 and n_2. The incident, reflected, and transmitted wavevectors are $\mathbf{k_i}$, $\mathbf{k_r}$ and $\mathbf{k_t}$, respectively.
   a) Use Snell's law to establish a relationship between the angles with respect to the z-axis, $\alpha_r = f(\alpha_i)$ and $\alpha_t = f(\alpha_i)$.
   b) Find an expression for α_i to achieve total integral reflection, and state the conditions for n_1 and n_2 for this to happen.
   c) If the incident wave is TM, find an expression for α_i that will yield 100% transmission (Brewster condition).
10. A stack of glass plates is placed in air (figure 2.16). Show that the emerging wave is parallel to the incident wave, no matter the incident angle and the refractive index and thickness of each slab.

Figure 2.17. Problem 12.

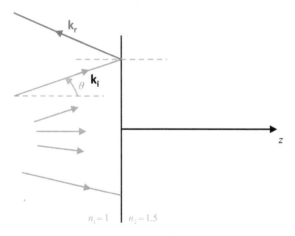

Figure 2.18. Problem 13.

11. Plot the (power) reflectivity and transmissivity coefficients for a plane wave travelling from air ($n = 1$) to glass (1.5) and from glass to air, for both TE and TM modes.

12. A plane wave travels in a waveguide with no power loss, i.e., under total internal reflection conditions (figure 2.17). The refractive indices are $n_1 = 1.55$, and $n_2 = 1.35$.
    a) Find the critical angle θ_c.
    b) Find the angle $\theta_{\pi/2}$ for which, under total internal reflection conditions, the phase shift between a TE and TM wave is $\pi/2$.
    c) For an incident angle $\theta = 60°$, how many reflections are necessary to achieve a phase shift between a TE and TM wave of $\pi/2$?

13. A beam, of divergence $\theta = 30°$ and intensity 1 W/srad is incident on a glass surface ($n = 1.5$) (figure 2.18). What is the total reflected intensity? Compare this value with the one when the incident field is a plane wave normal to the interface, of the same intensity.

14. Redo problem 13 when the beam is travelling from glass to air.

15. A point source emits light isotropically in water ($n = 1.33$). Plot the phase profile, $\phi(x, z = 0)$, of the field reflected by a flat interface with air on the other side (figure 2.19). Consider the TE and TM waves separately.

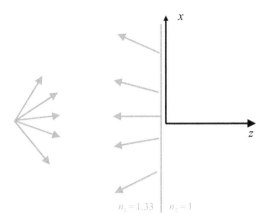

Figure 2.19. Problem 15.

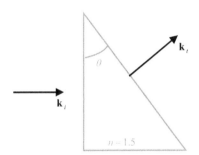

Figure 2.20. Problem 16.

16. A plane wave is incident on a prism of refractive index $n = 1.5$, as shown in figure 2.20. What percentage of the power is transmitted through the prism, if $\alpha = 30°$?

17. A beam of divergence $\theta = 15°$ is incident on the prism in figure 2.20. What is the divergence of the transmitted beam?

18. A plane wave of angular frequency ω propagates in a medium of conductivity σ. After what distance does the wave lose half of its power?

19. Write the scalar wave equation in the following representations
    a) $(x, y, k_z; t)$
    b) $(x, y, k_z; \omega)$
    c) (k_x, k_y, z, t)
    d) (k_x, k_y, z, ω)
    e) (x, k_y, k_z, t)
    f) (x, k_y, k_z, ω).

References and further reading

[1] Popescu G *Principles of Biophotonics, Volume 5—Light Propagation in Dispersive Media* (Bristol: IOP Publishing) (not yet published)

[2] Popescu G *Principles of Biophotonics, Volume 6—Light Propagation in Inhomogeneous Media* (Bristol: IOP Publishing) (not yet published)

[3] Popescu G *Principles of Biophotonics, Volume 7—Light Propagation in Nonlinear Media* (Bristol: IOP Publishing) (not yet published)

[4] Popescu G 2018 *Principles of Biophotonics, Volume 1—Linear Systems and the Fourier Transform in Optics* (Bristol: IOP Publishing) (not yet published)

[5] Bloembergen N 1965 *Nonlinear Optics: A Lecture Note and Reprint Volume. 3D Printing, With Addenda and Corrections* (Reading, MA: W. A. Benjamin) pp xviii, 229

[6] Göppert-Mayer M 1931 Über elementarakte mit zwei quantensprüngen *Ann. Phys.* **401** 273–94

[7] Popescu G *Principles of Biophotonics, Volume 9—Optical Imaging* (Bristol: IOP Publishing) (not yet published)

[8] Popescu G *Principles of Biophotonics, Volume 4—Light Propagation in Anisotropic Media* (Bristol: IOP Publishing) (not yet published)

[9] Fleisch D A 2008 *A Student's Guide to Maxwell's Equations* (Cambridge: Cambridge University Press)

[10] Kong J A 2002 *Maxwell Equations* (Cambridge, MA: EMW Publishing)

[11] Kong J A 2008 *Electromagnetic Wave Theory* (Cambridge, MA: EMW Publishing)

[12] Jin J 2010 *Theory and Computation of Electromagnetic Fields* (Hoboken, NJ: Wiley-IEEE Press)

[13] Orfanidis S J 2008 *Electromagnetic Waves and Antennas* (Piscataway, NJ: IEEE)

[14] Stratton J A 2007 IEEE Antennas and Propagation Society *Electromagnetic Theory* (Hoboken, NJ: Wiley)

[15] Yariv A and Yeh P 2007 *Photonics: Optical Electronics in Modern Communications* (New York: Oxford University Press)

Principles of Biophotonics, Volume 3
Field propagation in linear, homogeneous, dispersionless, isotropic media
Gabriel Popescu

Chapter 3

Propagation of electromagnetic fields

In this chapter, we describe *vector fields*, generated by anisotropic sources such as *electric* and *magnetic dipoles*. We first introduce a *vector Green's function* and study its relationship with the scalar Green's function.

3.1 Dyadic Green's function

Let us derive the solution (Green's function) of the vector wave equation when the source is an impulse (or a δ-function) in space. Recall Maxwell's equations in the space–frequency representation (section 2.3.1),

$$\nabla \times \mathbf{E}(\mathbf{r}, \omega) = i\omega\mu\mathbf{H}(\mathbf{r}, \omega) \tag{3.1a}$$

$$\nabla \times \mathbf{H}(\mathbf{r}, \omega) = -i\omega\varepsilon\mathbf{E}(\mathbf{r}, \omega) + \mathbf{j}(\mathbf{r}, \omega) \tag{3.1b}$$

$$\nabla \cdot \mathbf{D}(\mathbf{r}, \omega) = \rho(\mathbf{r}, \omega) \tag{3.1c}$$

$$\nabla \cdot \mathbf{B}(\mathbf{r}, \omega) = 0. \tag{3.1d}$$

We are interested in the fields generated by a certain current density, $\mathbf{j}(\mathbf{r}, \omega)$, in homogeneous and isotropic media, such as vacuum. Note that taking the divergence of equation (3.1b) and using equation (3.1c) yields the *continuity equation* in the $(\mathbf{r}, \omega)$ representation,

$$\begin{aligned} i\omega\rho(\mathbf{r}, \omega) &= \nabla \cdot \mathbf{j}(\mathbf{r}, \omega) \\ &= i\omega\varepsilon''\nabla \cdot \mathbf{E}. \end{aligned} \tag{3.2}$$

By eliminating $\mathbf{H}$ from equations (3.1a–b) and using $j = \sigma\mathbf{E} = \varepsilon''\omega\mathbf{E}$, we obtain the vector wave equation in vacuum

$$\nabla \times \nabla \times \mathbf{E}(\mathbf{r}, \omega) - \beta_0^2\mathbf{E}(\mathbf{r}, \omega) = i\omega\mu\mathbf{j}(\mathbf{r}, \omega),$$

$$\beta_0^2 = \frac{\omega^2}{c^2}. \tag{3.3}$$

We expand the first term in equation (3.3) using the property (see Volume 1, Appendix B [1] $\nabla \times \nabla \times \mathbf{E} = \bar{\bar{\nabla}}\nabla \cdot \mathbf{E} - \nabla^2\mathbf{E}$. The operator $\bar{\bar{\nabla}}\nabla$ is a dyadic operator,

defined as $\bar{\bar{\nabla}}\nabla = \begin{pmatrix} \dfrac{\partial}{\partial x^2} & \dfrac{\partial}{\partial x\partial y} & \dfrac{\partial}{\partial x\partial z} \\ \dfrac{\partial}{\partial y\partial x} & \dfrac{\partial}{\partial y^2} & \dfrac{\partial}{\partial y\partial zx} \\ \dfrac{\partial}{\partial z\partial x} & \dfrac{\partial}{\partial z\partial y} & \dfrac{\partial}{\partial z^2} \end{pmatrix}$. Throughout this book, we use the double

overbar, $\bar{\bar{A}}$, to denote such operators. Thus, the wave equation (equation (3.3)) becomes

$$\nabla^2\mathbf{E}(\mathbf{r}, \omega) + \beta_0^2\mathbf{E}(\mathbf{r}, \omega) = -i\omega\mu\mathbf{j}(\mathbf{r}, \omega) + \bar{\bar{\nabla}}\nabla\,\mathbf{E}$$
$$= -i\omega\mu\left[\mathbf{j}(\mathbf{r}, \omega) + \frac{1}{\omega^2\mu\varepsilon''}\bar{\bar{\nabla}}\nabla\mathbf{j}(\mathbf{r}, \omega)\right]$$
$$= -i\omega\mu\left[\bar{\bar{I}} + \frac{\bar{\bar{\nabla}}\nabla}{\beta_0^2}\right]\mathbf{j}(\mathbf{r}, \omega) \tag{3.4}$$

where $\bar{\bar{I}}$ is the unit operator, $\bar{\bar{I}} = \begin{pmatrix} 1 & 0 & 0 \\ 0 & 1 & 0 \\ 0 & 0 & 1 \end{pmatrix}$. In arriving at the final form of equation

(3.4), we used the continuity equation (equation (3.2)), i.e. $\nabla \cdot \mathbf{E} = \dfrac{1}{\omega\varepsilon''}\nabla \cdot \mathbf{j}(\mathbf{r}, \omega)$, and $\beta_0^2 = \omega^2/c^2 = \omega^2\varepsilon\mu$.

Next, we take the spatial Fourier transform with respect to $\mathbf{r}$. As usual, we keep the same symbols for the Fourier transforms, $\mathbf{E}$ and $\mathbf{j}$, but carry the arguments explicitly to avoid confusions, i.e., $\mathbf{E}(\mathbf{k}, \omega) \leftrightarrow \mathbf{E}(\mathbf{r}, t)$. Thus, equation (3.4). becomes

$$\left(\beta_0^2 - k^2\right)\mathbf{E}(\mathbf{k}, \omega) = -i\omega\mu\left[\bar{\bar{I}} - \frac{\bar{\bar{k}}k}{\beta_0^2}\right]\mathbf{j}(\mathbf{k}, \omega), \tag{3.5}$$

which gives the solution

$$\mathbf{E}(\mathbf{k}, \omega) = -\frac{i\omega\mu}{\beta_0^2 - k^2}\left[\bar{\bar{I}} - \frac{\bar{\bar{k}}k}{\beta_0^2}\right]\mathbf{j}(\mathbf{k}, \omega). \tag{3.6}$$

Equation (3.6) represents the full vector solution of the field generated by a current density $\mathbf{j}(\mathbf{k}, \omega)$. Note that $\bar{\bar{k}}k$ is the *dyadic* operator, Fourier-conjugate to $\bar{\bar{\nabla}}\nabla$, describing a second rank tensor, namely, $\bar{\bar{k}}k = \begin{pmatrix} k_x^2 & k_xk_y & k_xk_z \\ k_yk_x & k_y^2 & k_yk_z \\ k_zk_x & k_zk_y & k_z^2 \end{pmatrix}$.

The factor $\dfrac{-1}{\beta_0^{\,2} - k^2}$ is the *scalar* Green's function, $g(\mathbf{k}, \omega)$ associated with the 3D propagation, which is satisfied by each component of the E-field, namely,

$$\nabla^2 g(\mathbf{r}, \omega) + \beta_0^{\,2} g(\mathbf{r}, \omega) = -\delta(\mathbf{r})$$

$$g(\mathbf{k}, \omega) = -\frac{1}{\beta_0^{\,2} - k^2}. \tag{3.7}$$

Thus, equation (3.6) can be written in terms of this scalar Green's function,

$$\mathbf{E}(\mathbf{k}, \omega) = i\omega\mu \left[\bar{\bar{I}} - \frac{\bar{\mathbf{k}}\bar{\mathbf{k}}}{\beta_0^{\,2}} \right] g(\mathbf{k}, \omega)\mathbf{j}(\mathbf{k}, \omega)$$

$$= i\omega\mu G(\mathbf{k}, \omega)\mathbf{j}(\mathbf{k}, \omega), \tag{3.8}$$

where $\bar{\bar{G}}$ is a tensor given by

$$\bar{\bar{G}}(\mathbf{k}, \omega) = \left[\bar{\bar{I}} - \frac{\bar{\mathbf{k}}\bar{\mathbf{k}}}{\beta_0^{\,2}} \right] g(\mathbf{k}, \omega). \tag{3.9}$$

Writing explicitly the *dyadic operator*, $\bar{\bar{G}}$ can be expressed as

$$\bar{\bar{G}}(\mathbf{k}, \omega) = \frac{-1}{\beta_0^{\,2} - k^2} \begin{pmatrix} 1 - \dfrac{k_x^2}{\beta_0^{\,2}} & \dfrac{k_x k_y}{\beta_0^{\,2}} & \dfrac{k_x k_z}{\beta_0^{\,2}} \\[2mm] \dfrac{k_y k_x}{\beta_0^{\,2}} & 1 - \dfrac{k_y^2}{\beta_0^{\,2}} & \dfrac{k_y k_z}{\beta_0^{\,2}} \\[2mm] \dfrac{k_z k_x}{\beta_0^{\,2}} & \dfrac{k_z k_y}{\beta_0^{\,2}} & 1 - \dfrac{k_z^2}{\beta_0^{\,2}} \end{pmatrix}, \tag{3.10}$$

$$k^2 = k_x^2 + k_y^2 + k_z^2$$

Note that for an arbitrary current density source, $\mathbf{j}(\mathbf{k}, \omega) = \left[j_x(\mathbf{k}, \omega), j_y(\mathbf{k}, \omega), j_z(\mathbf{k}, \omega) \right]^T$ is simply obtained by multiplying it with the matrix in equation (3.10).

If the source current is an impulse of the form $\mathbf{j}(\mathbf{r}, \omega) = \mathbf{j}_0\delta(\mathbf{r})$, which implies $\mathbf{j}(\mathbf{k}, \omega) = \mathbf{j}_0$, the field solution is simply $\mathbf{E}(\mathbf{k}, \omega) \propto \bar{\bar{G}}(\mathbf{k}, \omega)$. Thus, $\bar{\bar{G}}(\mathbf{k}, \omega)$ is a *transfer function* and can be expressed as the Fourier transform of the *dyadic Green's function*, $\bar{\bar{G}}(\mathbf{r}, \omega)$,

$$\bar{\bar{G}}(\mathbf{r}, \omega) = \left[\bar{\bar{I}} + \frac{\bar{\nabla}\bar{\nabla}}{\beta_0^{\,2}} \right] g(\mathbf{r}, \omega). \tag{3.11}$$

Equation (3.11) relates the *dyadic* and the *scalar* Green's functions and is of great importance. $\bar{\bar{G}}(\mathbf{r}, \omega)$ generates the E-field solution for an arbitrary current density,

$\mathbf{j}(\mathbf{r}, \omega)$, via a 3D convolution integral. Thus, taking the Fourier transform of equation (3.8), we obtain

$$\mathbf{E}(\mathbf{r}, \omega) = i\omega\mu \iiint_V \bar{\bar{G}}(\mathbf{r} - \mathbf{r}', \omega)\mathbf{j}(\mathbf{r}', \omega)d^3\mathbf{r}'$$

$$= i\omega\mu\bar{\bar{G}}(\mathbf{r}, \omega)\otimes\mathbf{j}(\mathbf{r}, \omega). \tag{3.12}$$

In the following we employ this method of finding the E-field solution for two particular cases of broad interest, the *electric* and *magnetic* dipoles.

3.2 Electric dipole radiation

One of the most commonly used elementary source is the *electric dipole*, i.e., a pair of opposite sign charges $+q$, $-q$, separated by a distance, l,

$$\mathbf{p}(\mathbf{r}, t) = q(\mathbf{r}, t)\mathbf{l}. \tag{3.13}$$

The time change of the *dipole moment* is, by definition,

$$\frac{\partial\mathbf{p}(\mathbf{r}, t)}{\partial t} = \frac{\partial q(\mathbf{r}, t)}{\partial t}\mathbf{l}$$

$$= I(\mathbf{r}, t)\mathbf{l}, \tag{3.14}$$

where I is the current, defined as the flow of charge per unit time. Thus, in the $(\mathbf{r}, \omega)$ representation, the current dipole moment is

$$-i\omega\mathbf{p}(\mathbf{r}, \omega) = I(\mathbf{r}, \omega)\mathbf{l}. \tag{3.15}$$

Let us a consider an *elementary* current density along the z-axis, i.e., the source term in the wave equation of the form

$$\mathbf{j}(\mathbf{r}, \omega) = Il\delta(\mathbf{r})$$

$$= \hat{\mathbf{z}}Il\delta(\mathbf{r}) \tag{3.16}$$

where $\delta(\mathbf{r})$ is the usual 3D Dirac δ-function, which satisfies $\iiint_V \delta(\mathbf{r})d^3\mathbf{r} = 1$.

We now derive the dipole radiation E-field solution in the $(\mathbf{k}, \omega)$ representation, by using the dyadic Green's function derived previously (equation (3.10)), namely

$$\mathbf{E}(\mathbf{k}, \omega) = i\omega\mu\bar{\bar{G}}(\mathbf{k}, \omega)\mathbf{j}(\mathbf{k}, \omega) \tag{3.17a}$$

$$\bar{\bar{G}}(\mathbf{k}, \omega) = \frac{1}{\beta_0^2(k^2 - \beta_0^2)}\begin{pmatrix} \beta_0^2 - k_x^2 & k_xk_y & k_xk_z \\ k_yk_x & \beta_0^2 - k_y^2 & k_yk_z \\ k_zk_x & k_zk_y & \beta_0^2 - k_z^2 \end{pmatrix} \tag{3.17b}$$

$$\mathbf{j}(\mathbf{k}, \omega) = \begin{pmatrix} 0 \\ 0 \\ Il \end{pmatrix}. \tag{3.17c}$$

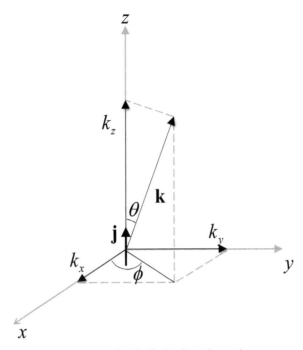

Figure 3.1. Electric dipole along the z-axis.

Thus, plugging equation (3.17b) and (3.17c) into (3.17a), we obtain

$$\mathbf{E}(\mathbf{k}, \omega) = \frac{i\omega\mu Il}{\beta_0^2(k^2 - \beta_0^2)}\begin{pmatrix} k_x k_z \\ k_y k_z \\ \beta_0^2 - k_z^2 \end{pmatrix}. \tag{3.18}$$

Equation (3.18) describes the full solution of the electric field vector radiated by an electric dipole, expressed in terms of the components of the wavevector $\mathbf{k}$ (see figure 3.1). The spatial dependence of $\mathbf{E}$ can be obtained by Fourier transforming back to space each component of $\mathbf{E}$.

Traditionally, the $\mathbf{E}$-field, $\mathbf{H}$-field, and Poynting vector for a dipole are expressed in cylindrical coordinates. Here we study the solution in equation (3.18), which has the advantage of being both compact and physically revealing, as follows, we notice that for $\mathbf{k} \| \mathbf{z}$, i.e. $k_x = k_y = 0$, $k = k_z$, the field components become $E_x = E_y = 0$, and $E_z(k) = $ const, meaning that the $\mathbf{E}$-field vanishes everywhere in space except at the origin, $E_z(\mathbf{r}) = \delta(\mathbf{r})$. This represents the well-known result that there is no field radiated along the direction of the dipole ($\hat{z}$ in this case), as illustrated in figure 3.2(a).

Another observation is that, for $\mathbf{k} \perp \hat{z}$, i.e. $k_z = 0$, that is, for propagation in a plane perpendicular to the dipole, the $\mathbf{E}$-field becomes

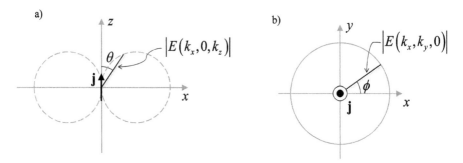

Figure 3.2. Amplitude of the electric field radiated by an electric dipole in the x–y (a) and x–y plane (b).

$$\mathbf{E}(k_x, k_y, 0, \omega) \propto \frac{1}{k_x^2 + k_y^2 - \beta_0^2}\begin{pmatrix} 0 \\ 0 \\ \beta_0^2 \end{pmatrix}, \tag{3.19}$$

thus, radiation perpendicular to the dipole is polarized along $\hat{z}$ and propagates with radial symmetry, i.e., $\mathbf{E}$ depends only on $k_\perp^2 = k_x^2 + k_y^2$ (figure 3.2(b)).

For propagation in the x–z plane, i.e., $k_y = 0$,

$$\mathbf{E}(k_x, 0, k_z, \omega) \propto \frac{1}{k_x^2 + k_z^2 - \beta_0^2}\begin{pmatrix} k_x k_z \\ 0 \\ \beta_0^2 - k_z^2 \end{pmatrix}. \tag{3.20}$$

Clearly, for propagation in the x–z plane, the radiation field pattern is no longer isotropic. Thus, using $k_x^2 + k_y^2 = \beta_0^2$, $k_x = \beta_0 \sin \theta$, $k_z = \beta_0 \cos \theta$, the magnitude squared of the electric field is

$$\begin{aligned}
\mid \mathbf{E}(k_x, 0, k_z, \omega) \mid^2 &= \frac{1}{(k_x^2 + k_z^2 - \beta_0^2)^2}\left[k_x^2 k_z^2 + (\beta_0^2 - k_z^2)^2 \right]. \\
&= \frac{1}{(k_\perp^2 - \beta_0^2)^2}\left[\beta_0^4 \sin^2 \theta \cos^2 \theta + \beta_0^4 (1 - \cos^2 \theta)^2 \right] \\
&= \frac{1}{(k_\perp^2 - \beta_0^2)^2}\left[\beta_0^4 \sin^2 \theta (\sin^2 \theta + \cos^2 \theta) \right] \\
&= \left(\frac{1}{k_\perp^2 - \beta_0^2}\beta_0^2 \sin \theta \right)^2.
\end{aligned} \tag{3.21}$$

Thus,

$$\mid \mathbf{E}(k_x, 0, k_z, \omega) \mid \propto \sin(\theta). \tag{3.22}$$

Equation (3.22) establishes a classical result, which states that in the plane of the dipole, the field amplitude is described by two tangent circles.

The magnetic field can be obtained immediately via the first Maxwell equation (Faraday's law), i.e.,

$$\mathbf{H}(\mathbf{k},\, \omega) = \frac{1}{\omega\mu}\mathbf{k} \times \mathbf{E}. \tag{3.23}$$

In the far field, the time average Poynting vector, which indicates the direction and magnitude of the power flow, has the form

$$\mathbf{S} = \frac{1}{2}\,\mathrm{Re}\,[\mathbf{E} \times \mathbf{H}^*]. \tag{3.24}$$

3.3 Magnetic dipole radiation

A *magnetic dipole* is a small current loop with an infinitesimally small radius. The current density is

$$\mathbf{j}(\mathbf{r},\, \omega) = \hat{\phi} I\delta(r_\perp - a)$$
$$= I\delta(r_\perp - a)\delta(z)\begin{pmatrix} -\sin\phi \\ \cos\phi \\ 0 \end{pmatrix}, \tag{3.25}$$

where $r_\perp = \sqrt{x^2 + y^2}$, $\phi = \arctan\left(\dfrac{y}{x}\right)$, with ϕ a time-dependent angle, indicating that the x- and y-components of the current are in quadrature ($\pi/2$ out of phase). Note that $\delta(r_\perp - a)$ indicates a ring of radius a in the x–y plane (figure 3.3). The 3D Fourier transform of $\mathbf{j}$, i.e., 1D transform with respect to z and Hankel transform with respect to $r_\perp$, gives (see volume 1, chapter 5 [1])

$$\mathbf{j}(\mathbf{k},\, \omega) = IJ_0(k_\perp a)\begin{pmatrix} -\sin\phi \\ \cos\phi \\ 0 \end{pmatrix}, \tag{3.26}$$

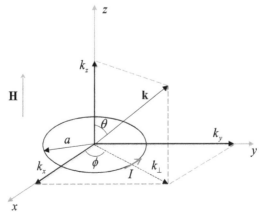

Figure 3.3. Magnetic dipole in the x–y plane.

where $k_\perp = \sqrt{k_x{}^2 + k_y{}^2}$ and J_0 is the Bessel function of the first kind and zeroth order.

Following the same procedure as for the electric dipole, specifically, using equations (3.17a–b), the resulting E-field is,

$$
\begin{aligned}
\mathbf{E}(\mathbf{k}, \omega) &= \frac{i\omega\mu I J_0(k_\perp a)}{\beta_0{}^2(k^2 - \beta_0{}^2)} \begin{pmatrix} \beta_0{}^2 - k_x{}^2 & k_x k_y & k_x k_z \\ k_y k_x & \beta_0{}^2 - k_y{}^2 & k_y k_z \\ k_z k_x & k_z k_y & \beta_0{}^2 - k_z{}^2 \end{pmatrix} \begin{pmatrix} -\sin\phi \\ \cos\phi \\ 0 \end{pmatrix} \\[2mm]
&= \frac{i\omega\mu I J_0(k_\perp a)}{\beta_0{}^2(k^2 - \beta_0{}^2)} \begin{pmatrix} -(\beta_0{}^2 - k_x{}^2)\sin\phi + k_x k_y \cos\phi \\ -k_x k_y \sin\phi + (\beta_0{}^2 - k_y{}^2)\cos\phi \\ 0 \end{pmatrix},
\end{aligned}
$$

(3.27)

where $k^2 = k_x{}^2 + k_y{}^2 + k_z{}^2 = k_\perp{}^2 + k_z{}^2$. Note that the z-component of E vanishes, $E_z = 0$. This is not surprising, as the loop generates a magnetic field along z instead.

For $\mathbf{k}\|\hat{z}$ $(k_x = k_y = 0)$, we find that equation (3.27) yields

$$
\begin{aligned}
\mathbf{E}(0,0,k_z,\omega) &= \frac{i\omega\mu I J_0(k_\perp a)}{\beta_0{}^2(k_z{}^2 - \beta_0{}^2)} \begin{pmatrix} -\beta_0{}^2 \sin\phi \\ \beta_0{}^2 \cos\phi \\ 0 \end{pmatrix} \\[2mm]
&= \frac{i\omega\mu I}{k_z{}^2 - \beta_0{}^2} \begin{pmatrix} -\sin\phi \\ \cos\phi \\ 0 \end{pmatrix}
\end{aligned}
$$

(3.28)

where we used that $J_0(0) = 1$. The solution indicates a *plane wave*, $\dfrac{1}{k_z{}^2 - \beta_0} \rightarrow \dfrac{e^{i\beta_0 z}}{\beta_0}$, which is *circularly polarized*, with the rotation in the direction of the current. Circular polarization is defined as two components of the electric field equal in amplitude and out of phase by $\pm\pi/2$, such as $(-\sin\phi, \cos\phi)$. A fuller description of polarization states will be given in volume 4 [2].

The derivations for the electric fields produced by the electric and magnetic dipoles highlight the efficiency of the calculations in the **k**-space. Following the same calculation procedure and the use of the dyadic transfer function, one can find solutions to more complex problems, as illustrated in the Problems section.

3.4 Problems

1. For the electric dipole described in the text, calculate the magnetic field **H**.
2. For the magnetic dipole described in the text, calculate the magnetic field **H**. Compare the expression for **E** of the electric dipole with **H** of the magnetic dipole.
3. Two electric dipoles, are separated by a distance d along the x-axis as shown in figure 3.4. The current density can be written as

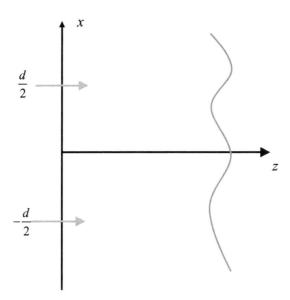

Figure 3.4. Problem 3.

$$\mathbf{j}(\mathbf{r},\,t) = j_0(t)\left[\delta\!\left(\mathbf{r} - \frac{d}{2}\hat{\mathbf{x}}\right) + \delta\!\left(\mathbf{r} + \frac{d}{2}\hat{\mathbf{x}}\right)\right]\hat{\mathbf{z}}$$

where $j_0(t)$ is the time dependence of the current density, assumed to be the same for both dipoles.

    (a) Calculate the electric field **E** emitted by the two dipoles.

    (b) Sketch the irradiance pattern in the x–y and x–z plane.

    (c) Calculate the pointing vector.

4. Redo problem 3 if one dipole is parallel to $\hat{z}$ and one parallel to $\hat{x}$, such that the current density reads (figure 3.5)

$$\mathbf{j}(\mathbf{r},\,t) = j_0(t)\left[\delta\!\left(\mathbf{r} - \frac{d}{2}\hat{\mathbf{x}}\right)\hat{\mathbf{z}} + \delta\!\left(\mathbf{r} + \frac{d}{2}\right)\hat{\mathbf{x}}\right]$$

5. The two dipoles in problem 3 are driven by the same current source, with the same frequency but a tunable phase shift, $\Delta\phi$. The current density can be written as

$$\mathbf{j}(\mathbf{r},\,t) = j_0\left[e^{i\omega_0 t}\delta\!\left(\mathbf{r} - \frac{d}{2}\hat{\mathbf{x}}\right) + e^{i(\omega_0 t + \Delta\phi)}\delta\!\left(\mathbf{r} + \frac{d}{2}\hat{\mathbf{x}}\right)\right]\hat{\mathbf{z}}$$

Calculate the electric field for $\Delta\phi = \pi/2,\ \pi$. Sketch the x–y and x–z irradiance contours.

6. Redo probe 5 when one dipole is along y and one along x.

7. Calculate the electric field emitted by an infinite array of electric dipoles, all parallel to z and of period d (figure 3.6). Discuss the effects of the dipole period d.

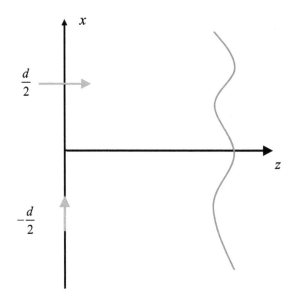

Figure 3.5. Problem 4.

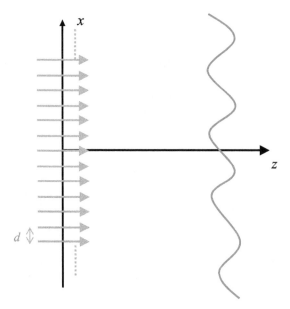

Figure 3.6. Problem 7.

8. Redo problem 7 if the dipoles are aligned along x.
9. A finite dipole array has a period d and aperture size L, as shown in figure 3.7. Calculate the electric field. Plot the irradiance contours in the x–y and x–z planes.
10. Redo problem 9 when the dipoles are along x.

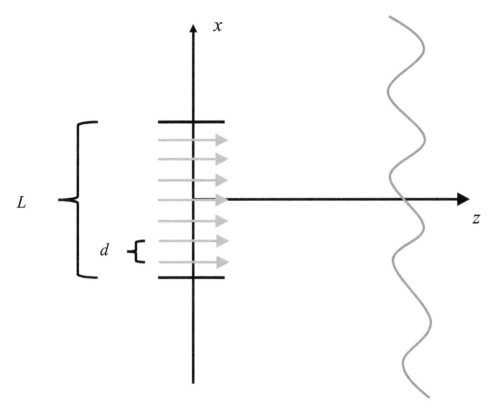

Figure 3.7. Problem 9.

11. Consider the infinite dipole array in problem 5. The dipoles are driven by a current source that applies a phase ramp across the array of the form $\phi(x) = \alpha x$. Calculate the electric field and discuss the differences with respect to the case of no phase ramp.

12. Redo problem 11 when the dipoles are along x.

13. Consider a finite phase array consisting of $2N + 1$ dipoles, equally spaced at period d, as shown in figure 3.8.

    The dipoles are gradually rotated by the same angle, such that the first and last dipole are antiparallel along z. Calculate the electric field and discuss the effect of the number of dipoles. Plot in the irradiance contours in the x–y and x–z planes.

14. Redo problem 13 when the first and last dipoles are antiparallel along x.

15. Redo problems 13 and 14, when the orientation difference between the first and last dipole is only 90°.

16. Calculate the electric field and plot the irradiance contours in the x–y and x–z planes for the four identical dipoles in figure 3.9 (consider current density $\mathbf{j}_0$).

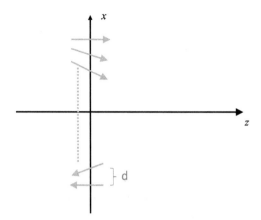

Figure 3.8. Problem 13.

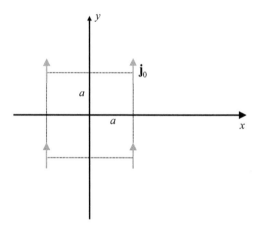

Figure 3.9. Problem 16.

17. Consider four identical dipoles oriented tangentially to a circle, as shown in figure 3.10. Calculate the electric field and plot the irradiance contours in the x–y and x–z planes.

18. Consider a 2D infinite dipole array, all parallel to x (figure 3.11). Calculate the electric field and the pointing vector. Plot $|\mathbf{E}|^2$ in the x–y and x–z planes.

19. Redo problem 18 if the orientation of the neighboring dipoles alternate between the $+\hat{\mathbf{x}}$ and $-\hat{\mathbf{x}}$ directions.

20. The dipoles in problem 18 are driven by a source that can introduce a phase delay of the form $\phi(x, y) = ax + by$. Calculate the radiated electric field and plot $|\mathbf{E}|^2$ in the x–y and x–z planes.

21. Redo problem 20 if the phase introduced is of the form $\phi(x, y) = \alpha(x^2 + y^2)$.

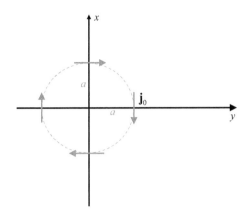

Figure 3.10. Problem 17.

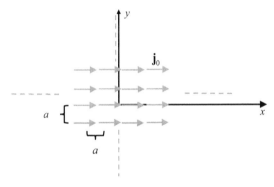

Figure 3.11. Problem 18.

22. Two magnetic dipoles lie in the x–z plane and are separated by a distance d along the x-axis, as shown in figure 3.12. Calculate the radiated electric field **E**.

23. Redo problem 22 if the current of the two magnetic dipoles are antiparallel.

24. The two currents in problem 22 are out of phase by $\Delta\phi$. Calculate the radiated electric field for $\Delta\phi = \pi/2, \ \pi$. Sketch the irradiance contour in the x–y and x–z planes.

25. Redo problem 22 if one dipole lies in the x–y and one in the x–z plane.

26. Calculate the electric field emitted by an infinite 1D array of magnetic dipoles in the x–z plane, distributed along the x-axis, with period a, as shown in figure 3.13. Discuss the effects of period a.

27. Redo problem 26 if the magnetic dipoles are oriented in the y–z plane, while still distributed along the x-axis.

28. A finite magnetic dipole array has a period a and aperture size L, as shown in figure 3.12. Calculate the electric field and plot the irradiance contours in the x–y and x–z plane (figure 3.14).

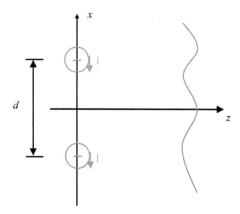

Figure 3.12. Problem 22.

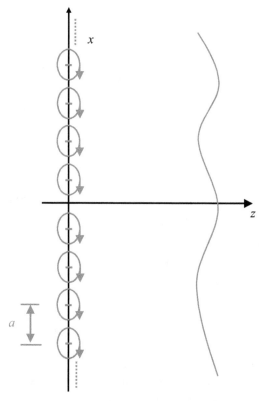

Figure 3.13. Problem 26.

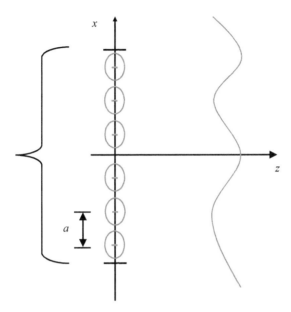

Figure 3.14. Problem 28.

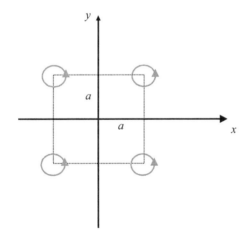

Figure 3.15. Problem 31.

29. Redo problem 28 when the magnetic dipoles are oriented parallel to the y–z plane, but remain distributed along the x-axis.
30. The infinite magnetic dipole array in problem 27 is driven by currents with a phase ramp across the array, $\phi(x) = \alpha x$. Calculate the electric field and discuss the differences with respect to the case in problem 27.
31. Calculate the electric field and plot the irradiance contours in the x–y and x–z planes for the four magnetic dipoles in figure 3.15.

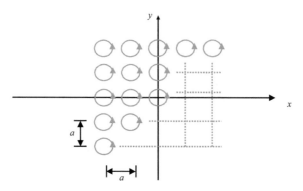

Figure 3.16. Problem 33.

32. Redo problem 31 if the currents of the magnetic dipoles along each diagonal are antiparallel.
33. Consider a 2D infinite magnetic dipole array, of period a, all lying in the x–y plane, as shown in figure 3.16. Calculate the electric field and plot the irradiance contours in the x–y and x–z planes.
34. Redo problem 33 if the direction of the current alternates between neighboring magnetic dipoles.
35. The magnetic dipoles in problem 33 are driven by a source that introduces a phase delay of the form $\phi(x, y) = ax + by$. Calculate **E** and plot $|E|^2$ in the x–y and x–z planes.
36. Redo problem 35 if $\phi(x, y) = \alpha(x^2 + y^2)$.

References and further reading

[1] Popescu G 2018 *Principles of Biophotonics, Volume 1—Linear Systems and the Fourier Transform in Optics* (Bristol: IOP Publishing) (not yet published)
[2] Popescu G *Principles of Biophotonics, Volume 4—Light Propagation in Anisotropic Media* (Bristol: IOP Publishing) (not yet published)
[3] Chew W C 1995 *Waves and Fields in Inhomogeneous Media* (New York: IEEE Press)
[4] Yariv A and Yeh P 2003 *Optical Waves in Crystals: Propagation and Control of Laser Radiation* (Hoboken NJ: Wiley)

IOP Publishing

Principles of Biophotonics, Volume 3
Field propagation in linear, homogeneous, dispersionless, isotropic media
Gabriel Popescu

Chapter 4

Propagation of scalar fields in free space

4.1 Primary and secondary sources

In this chapter we review the solutions of free space wave propagation. The Helmholtz equation (i.e., the wave equation in the $\mathbf{r} - \omega$ representation) that describes the propagation of a field emitted by a source s has the form

$$\nabla^2 U(\mathbf{r}, \omega) + \beta_0^2 U(\mathbf{r}, \omega) = s(\mathbf{r}, \omega). \tag{4.1}$$

In equation (4.1), U is the scalar field, which is a function of position $\mathbf{r}$ and frequency ω, β_0 is the vacuum wavenumber (or propagation constant), $\beta_0 = \omega/c$, and s is the source term, describing the spatial and spectral distribution of the source field. The source can be *primary* or *secondary*. A *primary source* is a *light-emitting* object. A *secondary source* is obtained by *re-radiation* of an illuminating field. For example, in volume 6 [1] we will study light scattering by inhomogeneous media, where the secondary source takes the form

$$s(\mathbf{r}, \omega) = U(\mathbf{r}, \omega)F(\mathbf{r}, \omega) \tag{4.2}$$

In equation (4.2), F is a function that describe the spatial distribution of the medium refractive index. Note that a secondary source exists as long as we provide an illumination field, while a primary source self-emits.

In order to solve equation (4.1) driven by a *primary* source, one needs to specify whether the propagation takes place in 1D, 2D, or 3D, and take advantage of any symmetry of the problem, e.g., spherical or cylindrical symmetry in 3D. Then, the *fundamental equation* associated with equation (4.1), which provides the Green's function of the problem, is obtained by replacing the source term with an impulse function, i.e., Dirac delta-function. The fundamental equation is then solved in the frequency domain, where the differential equation becomes algebraic. Below, we derive the well-known solutions for the propagation in 1D, 2D, and 3D.

4.2 1D Green's function: plane wave

For the 1D propagation (see figure 4.1), the fundamental equation is

$$\frac{\partial^2 g(x,\,\omega)}{\partial x^2} + \beta_0{}^2 g(x,\,\omega) = \delta(x), \qquad (4.3)$$

where g is Green's function, and $\delta(x)$ is the 1D delta-function, which in 1D describes a planar source of infinite size placed at the origin (figure 4.1(a)). More specifically, the source is constant (unity) and extends indefinitely in y and z, which can be written symbolically as $s(x, y, z) = \delta(x)1(y)1(z)$. Note that the source is also considered to be an impulse in time, $\delta(t)$, which yields a Fourier transform equal to 1 in the ω-domain.

Taking the Fourier transform with respect to x of equation (4.3) gives an algebraic equation

$$-k_x{}^2 g(k_x,\,\omega) + \beta_0{}^2 g(k_x,\,\omega) = 1, \qquad (4.4)$$

where we used the differentiation theorem of Fourier transforms, $\dfrac{d}{dx} \leftrightarrow ik_x$ (see volume 1 for details [2]). Thus, the frequency domain solution is simply

$$\begin{aligned}
g(k_x,\,\omega) &= -\frac{1}{k_x{}^2 - \beta_0{}^2} \\
&= \frac{-1}{2\beta_0}\left(\frac{1}{k_x - \beta_0} - \frac{1}{k_x + \beta_0} \right)
\end{aligned} \qquad (4.5)$$

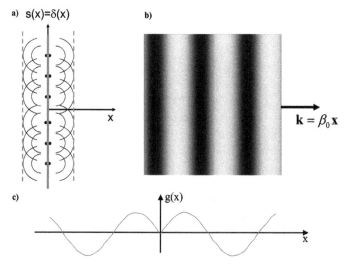

Figure 4.1. 1D propagation of the field from a planar source. (a) The source can be considered as a distribution of point sources. The superposition of all waves yield a plane wavefront. (b) The real part of the field propagating along x; **k** is the wave vector associated with the plane wave. (c) Green's function is real and even.

We can anticipate right away that, since $g(k_x)$ is real and even in k_x, its Fourier transform, i.e., Green's function $g(x)$ must be real and even as well (see chapter 7 in volume 1 [2]).

In order to find *Green's function*, i.e., the spatial-domain solution, we Fourier transform back $g(k_x, \omega)$ to the spatial domain. In taking the Fourier transform inverse of equation (4.5), we note that $1/(k_x - \beta_0)$ is the function $1/k_x$, shifted by β_0. Thus, we invoke the Fourier transform of $1/k_x$ and the shift theorem (see chapter 4 in volume 1 [2])

$$\frac{1}{k_x} \leftrightarrow i \, \text{sign}(x)$$

$$\frac{1}{k_0 - \beta_0} = i \, \text{sign}(x) e^{i\beta_0 x}, \tag{4.6}$$

$$\frac{1}{k_0 + \beta_0} = i \, \text{sign}(x) e^{-i\beta_0 x}$$

By combining equations (4.5) and (4.6), the solution through the 1D space, $x \in (-\infty, \infty)$, is

$$g(x, \omega) = \begin{vmatrix} -\dfrac{i}{2\beta_0}(e^{i\beta_0 x} - e^{-i\beta_0 x}), \text{ for } x \geqslant 0 \\[2mm] -\dfrac{i}{2\beta_0}(-e^{i\beta_0 x} + e^{-i\beta_0 x}), \text{ for } x < 0 \end{vmatrix}. \tag{4.7}$$

Simplifying equation (4.7), we obtain

$$g(x, \omega) = \begin{vmatrix} \dfrac{\sin \beta_0 x}{2\beta_0}, & x \geqslant 0 \\[2mm] -\dfrac{\sin \beta_0 x}{2\beta_0}, & x < 0 \end{vmatrix}, \tag{4.8a}$$

or,

$$g(x, \omega) = \frac{1}{2\beta_0} \sin \left| \beta_0 x \right|. \tag{4.8b}$$

As expected, we found that the Green's function is real and even. Figure 4.1 illustrates the emission from a planar source, which can be decomposed into point sources (figure 4.1(a)). The superposition of the emitted fields result in a plane wavefront propagating along a direction normal to the source (figure 4.1(b)). Figure 4.1(c) shows the x-dependence of Green's function (equation (4.8a)).

Function $g(x, \omega)$ consists of a superposition of two counter-propagating waves, of wavenumbers β_0 and $-\beta_0$. If instead we prefer to work with the *complex analytic signal* associated with $g(x, \omega)$, we simply suppress the negative frequency component. For a review of complex analytic signals, see chapter 7 in volume 1 [2]. In this case, the complex analytic solution is (we use the same notation, g)

$$g(x, \omega) = \frac{-i}{2\beta_0}e^{i\beta_0 \cdot x}, \quad x \geqslant 0. \tag{4.9}$$

Thus, we arrived at the well-known *plane wave* solution of the wave equation and established that it can be regarded as the complex analytic signal associated with the real field in equation (4.8). Note that since the wave equation contains second-order derivatives in space, $e^{ik_0 x}$, $e^{-ik_0 x}$ and their linear combinations are all valid solutions. The Green's function in equation (4.7), which we derived by solving the *fundamental equation* (equation (4.3)), is such a linear combination.

Thus, the amplitude of the plane wave is constant, but its phase is linearly increasing with the propagation distance (figure 4.2(a–b)). We note that, if the propagation direction is not parallel to one of the axes, the plane wave equation takes the general form

$$g(\mathbf{r}, \omega) = \frac{-i}{2\beta_0}e^{i\beta_0 \hat{\mathbf{k}} \cdot \mathbf{r}}, \tag{4.10}$$

where $\hat{\mathbf{k}}$ is the unit vector defining the direction of propagation, $\hat{\mathbf{k}} = \mathbf{k}/\beta_0$. Thus, the phase delay at a point described by position vector $\mathbf{r}$ is $\phi = k_{\|}r$, where $k_{\|}$ is the component of $\mathbf{k}$ parallel to $\mathbf{r}$ (figures (4.2c–d)).

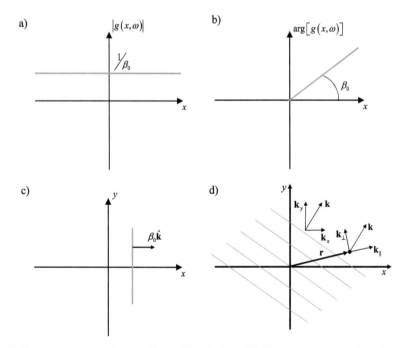

Figure 4.2. Plane wave propagation: amplitude (a) and phase (b). Plane wave propagation along x (c) and along an arbitrary direction (d).

4.3 2D Green's function: cylindrical wave

2D propagation happens when the source is an impulse in x–y, but is constant in one dimension. This source can be described as an infinite line source (figure 4.3). The source term can be described as $g(x, y, z) = \delta(x)\delta(y)1(z)$. The Green's function is obtained by solving the equation

$$\nabla^2 g(\mathbf{r}_\perp, \omega) + \beta_0^2 g(\mathbf{r}_\perp, \omega) = \delta(x)\delta(y)$$
$$\mathbf{r}_\perp = (x, y).$$
(4.11)

The Fourier transform of equation (4.11) gives ($\nabla \to i\mathbf{k}$)

$$\left(\beta_0^2 - k_\perp^2\right)g(k_\perp, \omega) = 1.$$
(4.12)

As expected, the solution of this equation only depends on the magnitude of the wavevector $|\mathbf{k}_\perp| = k_\perp$ and not its orientation, i.e., the propagation is isotropic in the x–y plane (figure 4.3).

The solution in the $\mathbf{k}$-domain (Fourier transform of Green's function, or the *transfer function*) is

$$g(\mathbf{k}_\perp, \omega) = \frac{1}{\beta_0^2 - k^2}$$
$$= \frac{1}{2k_\perp}\left[\frac{1}{\beta_0 - k_\perp} - \frac{1}{\beta_0 + k_\perp}\right].$$
(4.13)

Note that equation (4.13) is quite similar to equation (4.5), except that now the variable is $k_\perp = \sqrt{k_x^2 + k_y^2}$. In order to Fourier transform back this solution to space-domain, we use the Fourier transform of circularly-symmetric functions, i.e., the Hankel transform (see chapter 5, volume 1 [2]).

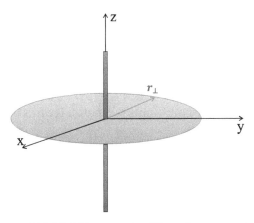

Figure 4.3. The line source yields 2D field propagation. The earliest wave arrives at moment $t = r_\perp/c$.

$$g(\mathbf{r}_\perp, \omega) = \frac{1}{2\pi} \int_0^\infty \frac{-1}{2k_\perp(k_\perp - \beta_0)} J_0(k_\perp r) k_\perp dk_\perp - \frac{1}{2\pi} \int_0^\infty \frac{1}{2k_\perp(k_\perp + \beta_0)} J_0(k_\perp r) k_\perp dk_\perp$$

$$= \frac{1}{4\pi} \int_0^\infty \frac{-1}{k_\perp - \beta_0} J_0(k_\perp r) dk_\perp - \frac{1}{4\pi} \int_0^\infty \frac{1}{k_\perp + \beta_0} J_0(k_\perp r) dk_\perp.$$

(4.14)

Note that, as described in chapter 5 of volume 1 [2], the shift theorem does not apply to Hankel transforms, thus, there is no simple evaluation of the intervals in equation (4.14).

However, an analytic form can be obtained in the time-domain. Since $\beta_0 = \omega/c$ and $d\omega = cd\beta_0$, we can take the inverse Fourier transform of equation (4.14) with respect to ω first and obtain the time-domain expression, namely,

$$g(\mathbf{r}_\perp, t) = \frac{1}{4\pi} \int_{-\infty}^\infty \int_0^\infty \frac{e^{-i\omega t}}{\frac{\omega}{c} - k_\perp} J_0(k_\perp r_\perp) dk_\perp d\omega - \frac{1}{4\pi} \int_{-\infty}^\infty \int_0^\infty \frac{e^{-i\omega t}}{\frac{\omega}{c} + k_\perp} J_0(k_\perp r_\perp) dk_\perp d\omega = \frac{c}{4\pi}$$

$$\int_0^\infty \int_{-\infty}^\infty \frac{e^{-i\omega t}}{\omega - ck_\perp} d\omega J_0(k_\perp r_\perp) dk_\perp - \frac{c}{4\pi} \int_{-\infty}^\infty \int_0^\infty \frac{e^{-i\omega t}}{\omega + ck_\perp} d\omega J_0(k_\perp r_\perp) dk_\perp,$$

(4.15)

where we reversed the order of integration. We see that the integral over ω can be easily evaluated using the Fourier transform of $1/\omega$ and the shift theorem, very similar to the calculation in section 4.2, namely,

$$\frac{1}{\omega} \leftrightarrow -i \, \text{sign}(t)$$

(4.16a)

$$\frac{1}{\omega \pm ck_\perp} \leftrightarrow -i \, \text{sign}(t) e^{\pm ik_\perp ct}$$

(4.16b)

Notice that there is a sign change between equation (4.16a) and the spatial frequency domain case, $1/k_x \leftrightarrow +i \, \text{sign}(x)$. Combining equation (4.16b) and (4.15), we obtain

$$g(\mathbf{r}_\perp, t) = \frac{-ic}{4\pi} \text{sign}(t) \left[\int_0^\infty J_0(k_\perp r_\perp) e^{-ik_\perp ct} dk_\perp - \int_0^\infty J_0(k_\perp r_\perp) e^{ik_\perp ct} dk_\perp \right]$$

(4.17)

In equation (4.17), we recognize the integrals as 1D Fourier transforms over $k_\perp$, with ct as the Fourier conjugate variable (units of length). It turns out that the 1D Fourier transform of J_0 is known as

$$J_0(k) \leftrightarrow \frac{\Pi\left(\frac{r}{2}\right)}{\pi\sqrt{1 - r^2}}, \qquad r \geqslant 0,$$

(4.18)

where $\Pi(r/2)$ is our usual 1D rectangular function,

$$\Pi\left(\frac{r}{2}\right) = \begin{vmatrix} 1, & \text{if } r \in [-1,1] \\ 0, & \text{rest.} \end{vmatrix} \tag{4.19}$$

Thus, plugging equation (4.18) into equation (4.17) and using the scaling property of the Fourier transform, we readily obtain

$$g(\mathbf{r}_\perp, t) = -\frac{ic \; \text{sign}(t)}{4\pi \; r_\perp} \left[\frac{\Pi\left(-\frac{ct}{2r_\perp}\right)| \, t < 0}{\pi\sqrt{1 - \left(\frac{ct}{r_\perp}\right)^2}} - \frac{\Pi\left(\frac{ct}{2r_\perp}\right)| \, t \geqslant 0}{\pi\sqrt{1 - \left(\frac{ct}{r_\perp}\right)^2}} \right] \tag{4.20}$$

We can re-write equation (4.20) by expanding $\text{sign}(t)$ and using that Π is an even function,

$$g(\mathbf{r}_\perp, t) = \frac{ic}{2\pi^2\sqrt{r_\perp^2 - (ct)^2}} \, \Pi\left(\frac{ct}{2r_\perp}\right)| \, t \geqslant 0. \tag{4.21}$$

Equation (4.21) can be expressed in a more physically intuitive way in terms of step functions, recalling that $\Pi(x/2b) = \Gamma(x + b) - \Gamma(x - b)$, namely,

$$g(\mathbf{r}_\perp, t) = \frac{i}{2\pi\sqrt{\frac{r_\perp^2}{c} - t^2}}\left[\Gamma\left(t + \frac{r_\perp}{c}\right) - \Gamma\left(t - \frac{r_\perp}{c}\right)\right]| \, t \geqslant 0$$

$$= \frac{1}{2\pi\sqrt{t^2 - \left(\frac{r_\perp}{c}\right)^2}}\left[\Gamma\left(t + \frac{r_\perp}{c}\right) - \Gamma\left(t - \frac{r_\perp}{c}\right)\right]| \, t \geqslant 0. \tag{4.22}$$

We see that the solution $g(\mathbf{r}_\perp, t)$ consists of two terms corresponding to an 'advanced' wave, which starts at an advanced time, $t + r_\perp/c$, and a 'retarded' one, of $t - r_\perp/c$. The latter term is the physically meaningful one, as the field at distance $r_\perp$ is delayed by $r_\perp/c$. Thus, the Green's function for a line source in free space is obtained by ignoring the first term,

$$g(\mathbf{r}_\perp, t) = \frac{1}{2\pi^2\sqrt{t^2 - \left(\frac{r_\perp}{c}\right)^2}}\Gamma\left(t - \frac{r_\perp}{c}\right). \tag{4.23}$$

4.4 3D Green's function: spherical wave

Let us obtain the Green's function associated with the wave propagation in a vacuum, or the response of free space to a point source (i.e., the impulse response). The fundamental equation becomes

$$\nabla^2 g(\mathbf{r}, \omega) + \beta_0{}^2 g(\mathbf{r}, \omega) = \delta^{(3)}(\mathbf{r})$$
$$\mathbf{r} = (x, y, z) \tag{4.24}$$
$$\beta_0 = \omega/c,$$

where $\delta^{(3)}$ represents a 3D delta-function. Like before, we Fourier transform equation (4.24) and use the relationship $\nabla \leftrightarrow i\mathbf{k}$ (see chapter 6, volume 1 [2]), which gives

$$-k^2 g(\mathbf{k}, \omega) + \beta_0{}^2 g(\mathbf{k}, \omega) = 1, \tag{4.25}$$

where $k^2 = \mathbf{k} \cdot \mathbf{k}$. Equation (4.25) readily yields the solution in the $(\mathbf{k}, \omega)$ representation

$$g(k, \omega) = \frac{1}{\beta_0{}^2 - k^2}, \tag{4.26}$$

where we see that g depends only on the magnitude of the wavevector, $k = | \mathbf{k} |$, and not its orientation, that is, the propagation is *isotropic*. Note that equation (4.26) looks quite similar to its analog in 1D (equation (4.5)) except that the x component of the wave vector k_x is now replaced by the magnitude of the wave vector, k. Thus, unlike in the 1D case, here we cannot apply the shift theorem directly.

In order to obtain the spatial domain solution, $g(\mathbf{r}, \omega)$, we now Fourier transform equation (4.26) back to the spatial domain. For propagation in isotropic media such as free space, the problem is spherically symmetric and, as shown in chapter 6, volume 1 [2], the 3D Fourier transform of equation (4.26) can be written in spherical coordinates as a 1D integral with sinc as the integral kernel,

$$g(r, \omega) = \int_0^\infty \frac{1}{k_0{}^2 - k^2} \frac{\sin(kr)}{kr} k^2 dk. \tag{4.27}$$

Using the partial fraction decomposition under the integrand and expressing $\sin(kr)$ in exponential form, we obtain

$$\begin{aligned}
g(\mathbf{r}, \omega) &= \frac{1}{2\pi} \int_0^\infty \frac{1}{2k} \left(\frac{1}{\beta_0 - k} - \frac{1}{\beta_0 + k} \right) \frac{e^{ikr} - e^{-ikr}}{2ikr} k^2 dk \\
&= \frac{1}{8\pi^2 ir} \left[\int_0^\infty \left(\frac{1}{\beta_0 - k} - \frac{1}{\beta_0 + k} \right) e^{ikr} d \right. \\
&\qquad k + \left. \int_0^\infty \left(\frac{1}{\beta_0 - (-k)} - \frac{1}{\beta_0 + (-k)} \right) e^{i(-k)r} dk \right] = \frac{1}{8\pi^2 ir} \\
&\int_{-\infty}^\infty \left(\frac{1}{\beta_0 - k} - \frac{1}{\beta_0 + k} \right) e^{ikr} dk.
\end{aligned} \tag{4.28}$$

In equation (4.28), we wrote the second term on the RHS as a function of $(-k)$ such that we can formally extend the limits of the integral from $-\infty$ to ∞ and obtain a 1D Fourier transform. This representation is only symbolic, as $k = | \mathbf{k} | \geqslant 0$, and we know from equation (4.26) that $g(k)$ is even.

We recognize the RHS of equation (4.28) breaks down into two 1D Fourier transforms, for which we can apply the shift theorem,

$$\int_0^\infty \frac{1}{k \mp \beta_0} e^{ikr} dk = i \, \text{sign}(r) e^{\pm i\beta_0 r}. \tag{4.29}$$

Thus, equation (4.29) simplifies to

$$\begin{aligned} g(\mathbf{r}, \omega) &= \frac{\text{sign}(r)}{8\pi^2 r} [e^{i\beta_0 r} - e^{-i\beta_0 r}] \\ &= \frac{\text{sign}(r)}{8\pi^2 r} 2i \sin \beta_0 r \\ &= \frac{i}{4\pi^2 r} \sin(|\beta_0 r|). \end{aligned} \tag{4.30}$$

Typically, we prefer to work with the complex analytic signal, which is obtained by ignoring the negative frequency term, $1/(k + \beta_0)$, or, equivalently, the 'incoming' wave $e^{-i\beta_0 r}$. Thus, we finally obtain the complex representation of the Green's function in 3D, namely the *spherical wave*,

$$g(\mathbf{r}, \omega) = \frac{e^{i\beta_0 r}}{r}, \quad r \geqslant 0, \tag{4.31}$$

where we ignored the irrelevant prefactor $i/4\pi^2$.

Not surprisingly, equation (4.31) resembles the 1D solution in section 4.1. However, there is a qualitative difference between the 1D and 3D cases, namely, the $1/r$ amplitude decay (figure 4.4). Knowing the Green's function associated with free space (equation (4.31)), i.e., the response to a point source, we can easily calculate the response to an arbitrary source $s(\mathbf{r}, \omega)$ via a *convolution integral*,

$$\nabla^2 U(\mathbf{r}, \omega) + k_0^2 U(\mathbf{r}, \omega) = s(\mathbf{r}, \omega)$$

$$U(\mathbf{r}, \omega) = \int_V s(\mathbf{r}', \omega) \frac{e^{ik_0|\mathbf{r}-\mathbf{r}'|}}{|\mathbf{r} - \mathbf{r}'|} d^3\mathbf{r}', \tag{4.32}$$

where the integral is performed over the volume of the source.

Equation (4.32) is the essence of *Huygens principle* (seventeenth century), which establishes that, upon propagation, points set in oscillation by the field become new sources, and the emerging field is the summation of spherical wavelets emitted by all these point sources. Thus, each point reached by the field becomes a secondary

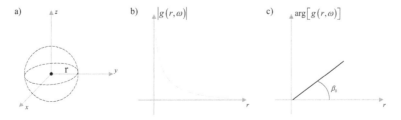

Figure 4.4. (a) Propagation of spherical waves. (b) Amplitude versus r. (c) Phase versus r.

source, which emits a new spherical wavelet and so on. Note that Huygens was able to propose a correct picture for wave propagation almost two centuries before Maxwell's electromagnetic theory. Huygens assumed that each secondary source emits light only in the forward semi-space and he could not explain diffraction. Later, Fresnel placed Huygens' principle on a more solid mathematical background, which also explained diffraction phenomena. We will discuss this phenomenon in detail, in chapter 5.

4.5 Problems

1. An infinite planar source is positioned at $x = a$ and emits monochromatic light at frequency ω_0, as shown in figure 4.5. Write down the expression of the field in free space.
2. An infinite plane source makes an angle θ with the x-axis and emits light at frequency ω_0 in free space. Calculate the emitted field (figure 4.6).
3. A pair of infinite plane sources emit light of frequency ω_0. The two sources are separated by a distance $2a$ (figure 4.7). Calculate the expression of the field between the two sources.
4. Two planar sources make angle θ_1 and θ_2 with respect to the x-axis, as shown in figure 4.8. The two sources are separated by a distance $2a$ and emit light at frequency ω_0. Calculate the field between the two sources.
5. Redo problems 3 and 4 when the two sources emit frequencies ω_1 and ω_2. A detector is placed at position $x = b$ between the two sources. Calculate the temporal-averaged irradiance detected when the detector integration time τ_0 satisfies

   a. $\tau_0 \ll \dfrac{2\pi}{|\omega_2 - \omega_1|}$

   b. $\tau_0 \gg \dfrac{2\pi}{|\omega_2 - \omega_1|}$.

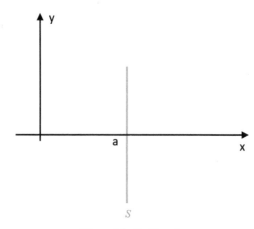

Figure 4.5. Problem 1.

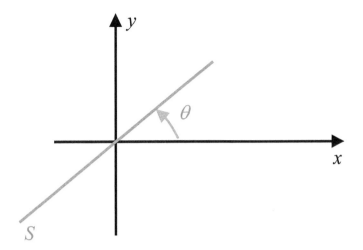

Figure 4.6. Problem 2.

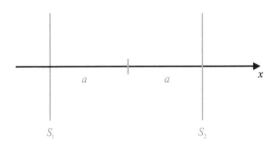

Figure 4.7. Problem 3.

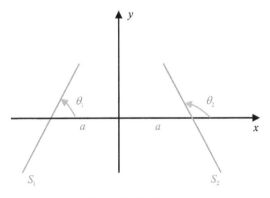

Figure 4.8. Problem 4.

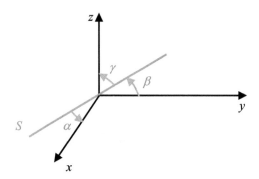

Figure 4.9. Problem 6.

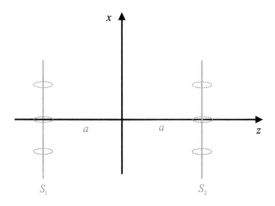

Figure 4.10. Problem 7.

6. An infinite line source emits monochromatic light, as shown in figure 4.9. Calculate the field emitted in free space.

7. Two infinite wires are parallel and separated by a distance $2a$. Calculate the total field propagating in free space (figure 4.10).

8. A point source is placed at the position $\mathbf{r}_0 = (x_0, 0, z_0)$. Calculate the field at the plane $z = 0$, if the optical frequency is ω_0 (figure 4.11).

9. A point source is at a distance a from a flat mirror. In order to calculate the resulting field propagating in free space, one can consider that the reflected field originates at the image of the source as formed by the mirror (figure 4.12). Calculate the total field in free space (assume 100% reflectivity and frequency ω_0).

10. Two point sources are separated by a distance $2a$, as shown in figure 4.13. Calculate the total field in free space, if the two sources emit at frequency ω_0.

11. An infinite array of point sources are uniformly distributed along the x-axis with a period a. Calculate the total field emitted, if all sources emit at ω_0.

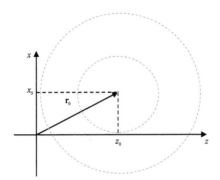

Figure 4.11. Problem 8.

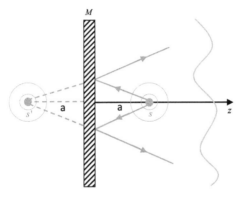

Figure 4.12. Problem 9. s' is the image of point source s. M is a perfect mirror.

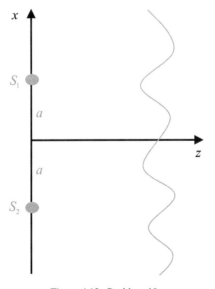

Figure 4.13. Problem 10.

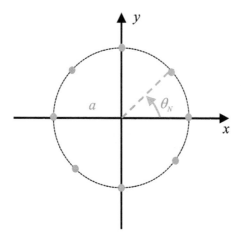

Figure 4.14. problem 13.

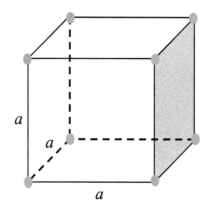

Figure 4.15. Problem 15.

12. An infinite 2Darray of point sources are uniformly distributed in $x-y$ with a period a in both directions. Calculate the emitted field, if all sources emit at ω_0.

13. An array of $2N$ point sources are uniformly distributed along a circle of radius a. Calculate the field emitted at ω_0. Discus the case $N \rightarrow \infty$ (figure 4.14).

14. An infinite 3D array of point sources are distributed in x, y, and z with periods a, b, and c, respectively. Calculate the total field emitted at frequency ω_0.

15. Calculate the field emitted by eight point sources placed at the corners of a cube of radius a (figure 4.15).

16. Calculate the total field in free space emitted by an infinite wire that is parallel to a perfect mirror, at a distance a from it (figure 4.16).

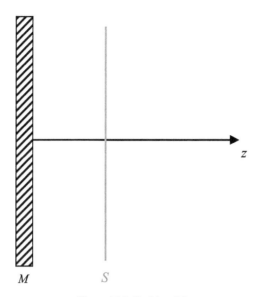

Figure 4.16. Problem 16.

References and further reading

[1] Popescu G *Principles of Biophotonics, Volume 6—Light Propagation in Inhomogeneous Media* (Bristol: IOP Publishing) (not yet published)

[2] Popescu G 2018 *Principles of Biophotonics, Volume 1—Linear Systems and the Fourier Transform in Optics* (Bristol: IOP Publishing)

[3] Evans G, Blackledge J M and Yardley P 2000 *Analytic Methods for Partial Differential Equations* (London; New York: Springer)

[4] Born M and Wolf E 1999 *Principles of Optics: Electromagnetic Theory of Propagation, Interference and Diffraction of Light* (Cambridge: Cambridge University Press)

[5] Chew W C 1995 *Waves and Fields in Inhomogeneous Media* (New York: IEEE Press)

IOP Publishing

Principles of Biophotonics, Volume 3
Field propagation in linear, homogeneous, dispersionless, isotropic media
Gabriel Popescu

Chapter 5

Diffraction of scalar fields

5.1 Diffraction by a 2D object

Sommerfeld described 'diffraction' as any deviation of light rays from rectilinear path which cannot be interpreted as reflection or refraction [1]. In this book, we consider diffraction the interaction of light with 2D objects, such as apertures, gratings, transparent screens, etc. Interaction with 3D objects falls under 'light scattering' and will be discussed separately in detail in volume 6 [2].

Any optical instrument is bound to contain certain apertures somewhere in the optical path. Even in the absence of irises or pinholes purposely inserted in the system, optical components, such as lenses, are typically held by mounts that themselves can create diffraction. In this chapter, we develop the mathematical apparatus to describe this phenomenon and study various approximations that can simplify the description.

According to Born and Wolf, 'Diffraction problems are amongst the most difficult ones encountered in optics' [3]. During a period when Newton's corpuscular theory of light [4] was generally accepted, Huygens' incredible intuition provided new insights into the propagation of optical fields (figure 5.1 and reference [5]). Huygens' principle states that points on the wavefront become secondary sources that emit spherical waves. The subsequent wavefronts are the result of superposition of these secondary waves (figure 5.1). Later on, as Young [6] and Fresnel [7] cemented the wave theory of light, Huygens' principle was placed in a more rigorous mathematical formalism (for a collections of memoirs by Huygens, Young, and Fresnel, one can consult reference [8]). Informative reviews in historical context of the scalar diffraction theory can be found in [3] (chapter VII) and [9] (chapter 3).

After the discovery of Maxwell's equations, Kirchhoff and Sommerfeld developed diffraction theories that are rooted in the wave equation. Goodman gave a historical account of these developments (see chapter 3 in reference [9]). Here, we describe diffraction by taking advantage of the linear systems approach and Green's functions derived earlier.

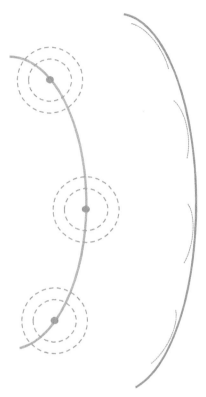

Figure 5.1. Illustration of Huygens' principle: points on the wavefront become secondary sources that emit spherical waves. The subsequent wavefronts are the result of superposition of these secondary waves.

Let us consider the geometry in figure 5.2, where an incident field, U_i, is incident on a thin screen, characterized by a 2D transmission function $t(\mathbf{r}_\perp)$. The transmission function is generally complex, $t(\mathbf{r}_\perp) = |\,t(\mathbf{r}_\perp)|\,e^{i\phi(\mathbf{r}_\perp)}$, and can describe various objects, for example

$$t(\mathbf{r}_\perp) = \Pi\!\left(\frac{\mathbf{r}_\perp}{2a}\right), \text{ circular aperture of radius } a \tag{5.1a}$$

$$t(\mathbf{r}_\perp) = e^{i\phi(\mathbf{r}_\perp)}, \text{ phase screen} \tag{5.1b}$$

$$t(\mathbf{r}_\perp) = \Pi\!\left(\frac{x}{2a}\right)\!\circledcirc \mathrm{comb}\!\left(\frac{x}{2b}\right), \text{ 1D amplitude grating} \tag{5.1c}$$

In order to find the diffracted field emerging from such a 2D object, we start with the inhomogeneous Helmholtz equation, containing a position dependent, complex refractive index, $n(\mathbf{r}) = n'(\mathbf{r}) + in''(\mathbf{r})$

$$\nabla^2 U(\mathbf{r}, \omega) + n^2(\mathbf{r})\beta_0^2 U(r, \omega) = 0 \tag{5.2}$$

In section 4.4, we solved the homogenous version of equation (5.2) ($n = 1$) and found that its Green's function is the well-known spherical wave. Therefore, it is

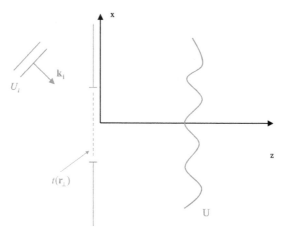

Figure 5.2. Diffraction by a 2D object characterized by a transmission function $t(x,|, y)$. U_i is the incident field, assumed to be a plane wave of wavevector $\mathbf{k_i}$ and U is the diffracted field of interest.

helpful to separate equation (5.2) into its *homogenous* and *inhomogeneous* parts, as follows,

$$\nabla^2 U(\mathbf{r}, \omega) + \beta_0^2 U(r, \omega) = -\beta_0^2[n^2(\mathbf{r}) - 1]U(r, \omega) \tag{5.3}$$

In order to specify the 2D nature of the object, we consider it as being a thin slice in z

$$[n^2[(\mathbf{r}) - 1]\Pi\left(\frac{z}{\Delta z}\right) \simeq [n^2(\mathbf{r_\perp}) - 1]\Delta z\delta(z)], \tag{5.4}$$

where we used the fact that a rectangular function normalized to unit area approaches a δ-function as the thickness approaches zero. We note that $\beta_0[n^2(\mathbf{r_\perp}) - 1]\Delta z$ is a unitless object property, which describes its complex transmission function

$$\begin{aligned} t(\mathbf{r_\perp}) &= \beta_0[n^2(\mathbf{r_\perp}) - 1]\Delta z \\ &= |\, t(\mathbf{r_\perp})|\, e^{i\phi(\mathbf{r_\perp})} \end{aligned} \tag{5.5}$$

Next, we consider that, at the plane of the screen, $z = 0$, the field is approximated well by the incident field. Thus, equation (5.3) becomes

$$\nabla^2 U(\mathbf{r}, \omega) + \beta_0^2 U(r, \omega) = -\beta_0 t(\mathbf{r_\perp})\delta(z)U_i(\mathbf{r}, \omega) \tag{5.6}$$

Equation (5.6) can be solved easily in the $\mathbf{k}$ domain. However, we know that the solution of equation (5.6) is nothing but the convolution between the source term (the RHS) and the spherical wave, namely

$$\begin{aligned} U(\mathbf{r}, \omega) &= [\beta_0 t(\mathbf{r_\perp})\delta(z)U_i(\mathbf{r}, \omega)]\circledast_r \frac{e^{i\beta_0 r}}{r} \\ &= \beta_0 \int t(\mathbf{r_\perp'})\delta(z')U_i(\mathbf{r'}, \omega)\frac{e^{i\beta_0|\mathbf{r}-\mathbf{r'}|}}{|\,\mathbf{r} - \mathbf{r'}\,|}d^3\mathbf{r'} \end{aligned} \tag{5.7}$$

where $| \mathbf{r} - \mathbf{r}' | = \sqrt{(x - x'^2) + (y - y')^2 + (z - z')^2}$.

The effect of $\delta(z')$ is to set $z' = 0$ under the integral and, thus, reduce the convolution to only two dimensions,

$$U(\mathbf{r}, \omega) = \beta_0[t(\mathbf{r}_\perp) U_i(\mathbf{r}_\perp, 0, \omega)] \otimes_{r_\perp} \frac{e^{i\beta_0 r}}{r}, \tag{5.8}$$

Equation (5.8) provides a general solution and physical insight into the diffraction phenomenon. Let us consider that the incident field is a plane wave propagating along the z-direction,

$$U_i(\mathbf{r}, \omega) = A(\omega)e^{i\beta_0 z}, \tag{5.9}$$

where, $A(\omega)$ is the spectral amplitude of the field. Plugging equation (5.9) into (5.8), we obtain right away the simple solution

$$U(\mathbf{r}, \omega) = \beta_0 A(\omega)e^{i\beta_0 z}t(\mathbf{r}_\perp) \otimes_{r_\perp} \frac{e^{i\beta_0 r}}{r} \tag{5.10}$$

Equation (5.10) indicates that, if the plane $z = 0$ is uniformly illuminated, the solution is nothing more than summing the spherical waves originating at each point in the plane (figure 5.3). This phenomenon is fully consistent with Huygens' principle and highlights his extraordinary insight, more than three centuries ago.

Let us consider now that the incident field is propagating along an arbitrary direction, as shown in figure 5.2,

$$U_i(\mathbf{r}, \omega) = A(\omega)e^{i\mathbf{k}_i \cdot \mathbf{r}}. \tag{5.11}$$

The diffracted field becomes

$$U(\mathbf{r}, \omega) = \beta_0 A(\omega)[t(\mathbf{r}_\perp)e^{i\mathbf{k}_{i\perp} \cdot \mathbf{r}_\perp}] \otimes_{\mathbf{r}_\perp} \frac{e^{i\beta_0 r}}{r}. \tag{5.12}$$

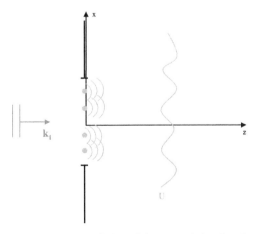

Figure 5.3. Diffraction by a screen as a convolution of the transmission function with the spherical wave, which is a manifestation of Huygens' principle.

In essence, the factor $e^{i\mathbf{k}_{i\perp}\cdot\mathbf{r}_\perp}$ accounts for the phase delay between various points in the plane $z = 0$ that the tilted incident wave introduces. Interestingly, in the frequency domain, we see that the tilt adds a spatial frequency shift, according to the shift theorem.

$$t(\mathbf{r}_\perp)e^{i\mathbf{k}_{i\perp}\cdot\mathbf{r}_\perp} \quad\leftrightarrow\quad t(\mathbf{k}_\perp - \mathbf{k}_{i\perp}) \tag{5.13}$$

Equation (5.13) indicates that tilting the incident field shifts the frequency content of the transmission function to higher values compared to that of on-axis illumination. This is a practical way to encode the finer structure of the 2D object of interest into the diffracted field and is often exploited in the field of biophotonics.

The convolution with the spherical wave in equation (5.12) is difficult to calculate analytically mainly because of the presence of the squared root in evaluating the distance $|\mathbf{r} - \mathbf{r}_i|$. In the following sections, we describe this operation in a different variable representation and also provide several levels of approximations for the spherical wave, which simplifies the calculations.

5.2 Plane wave decomposition of spherical waves: Weyl's formula

In diffraction problems, we often aim to derive an expression for the field at a plane perpendicular the optical axis, at a certain distance z from the object. It turns out that a spherical wave can be expressed in the $(k_\perp, z)$ representation, in analytic form. Recalling the $\mathbf{k}$-vector representation of the 3D Green function, we have (see section 4.4)

$$
\begin{aligned}
g(\mathbf{k}, \omega) &= \frac{-1}{k_x^2 + k_y^2 + k_z^2 - \beta_0^2} \\
&= \frac{-1}{k_z^2 - \gamma^2(\mathbf{k}_\perp)}
\end{aligned}
\tag{5.14}
$$

where $\gamma^2(\mathbf{k}_\perp) = \beta_0^2 - k_\perp^2$. In order to take the Fourier transform inverse of $g(\mathbf{k}, \omega)$ with respect to k_z, we use the usual partial fraction decomposition,

$$g(\mathbf{k}, \omega) = -\frac{1}{2\gamma(\mathbf{k}_\perp)}\left[\frac{1}{k_z - \gamma(\mathbf{k}_\perp)} - \frac{1}{k_z + \gamma(\mathbf{k}_\perp)}\right]. \tag{5.15}$$

Note the similarity between equation (5.15) and equation (4.5) in section 4.2, describing the 1D propagation. Similar to equation (4.6), we can take the Fourier transform inverse to the z-domain as

$$\frac{1}{k_z - \gamma(\mathbf{k}_\perp)} \leftrightarrow i\ sign(z)e^{i\gamma(\mathbf{k}_\perp)z} \tag{5.16a}$$

$$\frac{1}{k_z + \gamma(\mathbf{k}_\perp)} \leftrightarrow i\ sign(z)e^{-i\gamma(\mathbf{k}_\perp)z} \tag{5.16b}$$

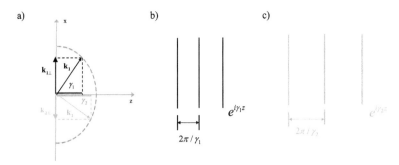

Figure 5.4. (a) Plane wave decomposition of a spherical wave: $\mathbf{k}_1$ and $\mathbf{k}_2$ are arbitrary wavevectors, $|\mathbf{k}_1| = |\mathbf{k}_2| = \beta_0$. Their respective projections on z are γ_1, and γ_1. (b) Plane wave propagation along z for $\gamma_1(\mathbf{k}_{1\perp})$. The period is $2\pi/\gamma_1$. (c) Plane wave propagation along z for $\gamma_2(\mathbf{k}_{2\perp})$. The period is $2\pi/\gamma_2$. Since $\gamma_2 > \gamma_1$ the period $2\pi/\gamma_2$ is smaller.

Choosing the complex analytic signal and only the outgoing wave (equation (5.15a)), we obtain

$$g(\mathbf{k}_\perp, z; \omega) = i\frac{e^{i\gamma(\mathbf{k}_\perp)z}}{2\gamma(\mathbf{k}_\perp)}. \tag{5.17}$$

We see that $g(\mathbf{k}_\perp, z)$ is just the Fourier transform of the spherical wave over $\mathbf{r}_\perp$, or

$$\frac{e^{i\beta_0 r}}{r} = i\iint \frac{e^{i\gamma(\mathbf{k}_\perp)z}}{2\gamma(\mathbf{k}_\perp)}\, e^{i\mathbf{k}_\perp\cdot\mathbf{r}_\perp}d^2\mathbf{k}_\perp.$$

The solution $g(k_\perp, z)$ is very similar to the 1D plane wave solution, except β_0 is now replaced by $\gamma(\mathbf{k}_\perp)$ and there is also an amplitude term, $1/\gamma$. Thus, $g(\mathbf{k}_\perp, z)$ can be regarded as a collection of plane waves, each propagating with a $\mathbf{k}_\perp$-dependent wavenumber, γ (see figure 5.4).

For this reason, equation (5.16) is known as the *plane wave decomposition*. This formula was derived first by Weyl in 1919 [10].

The $(k_\perp, z)$ representation provides an efficient way for calculating the diffracted field discussed in section 5.1. Let us take the Fourier transform (F.T.) with respect to $\mathbf{r}$ of the Helmholtz equation (equation (5.5)) to retrieve $U(\mathbf{k}, \omega)$,

$$U(\mathbf{k}, \omega) = \frac{\beta_0}{k^2 - \beta_0^2}[t(\mathbf{k}_\perp)\otimes_\mathbf{k} U_i(\mathbf{k})], \tag{5.18}$$

where the convolution over k_z yields a simple integral of U_i over k_z, as $t(\mathbf{k}_\perp)$ has no k_z dependence. Taking the F.T. inverse with respect to k_z, we use the fact that the term $1/(k^2 - \beta_0^2)$ yields the result in equation (5.16),

$$\begin{aligned}U(\mathbf{k}_\perp, z; \omega) &= i\beta_0\frac{e^{i\gamma(\mathbf{k}_\perp)z}}{2\gamma(\mathbf{k}_\perp)}\otimes_z[t(\mathbf{k}_\perp)\delta(z)\otimes_{\mathbf{k}_\perp} U_i(\mathbf{k}_\perp, z)]\\ &= i\beta_0\frac{e^{i\gamma(\mathbf{k}_\perp)z}}{2\gamma(\mathbf{k}_\perp)}\otimes_z[\delta(z)t(\mathbf{k}_\perp)\otimes_{\mathbf{k}_\perp} U_i(\mathbf{k}_\perp, 0)]\end{aligned} \tag{5.19}$$

where we used the property of the delta-function that yields $\delta(z)U_i(\mathbf{k}_\perp, z) = U_i(\mathbf{k}_\perp, 0)\delta(z)$. Finally, equation (5.19) simplifies further as one term if the convolution over z is $\delta(z)$, and we know the property that any function convolved with a delta-function returns the function itself, namely $e^{i\gamma(\mathbf{k}_\perp)z}\otimes_z\delta(z) = e^{i\gamma(\mathbf{k}_\perp)z}$,

$$U(\mathbf{k}_\perp, z; \omega) = i\beta_0[t(\mathbf{k}_\perp)\otimes_{\mathbf{k}_\perp} U_i(\mathbf{k}_\perp, 0)]\frac{e^{i\gamma(\mathbf{k}_\perp)z}}{2\gamma(\mathbf{k}_\perp)}. \tag{5.20}$$

Equation (5.20) exhibits an interesting feature in that the only z- dependence comes from the phase term $e^{i\gamma(\mathbf{k}_\perp)z}$, which makes predicting the diffracted field distribution at various planes $z = z_1, z_2$, etc, particularly easy. The result simplifies further if the incident field is a plane wave, say $U_i(\mathbf{r}, \omega) = A(\omega)e^{i\mathbf{k}_i\cdot\mathbf{r}}$, which in the $(\mathbf{k}_\perp, z)$ domain gives

$$U_i(\mathbf{k}_\perp, z; \omega) = A(\omega)\delta(\mathbf{k}_\perp - \mathbf{k}_{i\perp})e^{ik_{iz}z}. \tag{5.21}$$

Combining equation (5.20) and equation (5.21), we arrive at

$$U(\mathbf{k}_\perp, z; \omega) = i\beta_0 A(\omega)t(\mathbf{k}_\perp - \mathbf{k}_{i\perp})\frac{e^{i\gamma(\mathbf{k}_\perp)z}}{2\gamma(\mathbf{k}_\perp)} \tag{5.22}$$

Of course, if U_i propagates along z, $\mathbf{k}_{i\perp} = 0$ and the diffracted field is

$$U(\mathbf{k}_\perp, z; \omega) = i\beta_0 A(\omega)t(\mathbf{k}_\perp)\frac{e^{i\gamma(\mathbf{k}_\perp)z}}{2\gamma(\mathbf{k}_\perp)}. \tag{5.23}$$

Equation (5.23) is encountered in practice very often, as it can describe the diffraction at a screen under normal incidence.

Importantly, if we know the field at a particular plane z_1, the field at z_2 can be simply calculated by multiplying with $e^{i\gamma(\mathbf{k}_\perp)(z_2-z_1)}$. Another interesting feature is that $|U(\mathbf{k}_\perp, z; \omega)|^2$ is independent of z. In other words, the spatial power spectrum is z-invariant. This fact can be understood by stating that the angular spectrum of the light is defined at the plane of the object (see figure 5.5), i.e., the distribution of transverse $\mathbf{k}$-vectors, $\mathbf{k}_\perp$, stay the same.

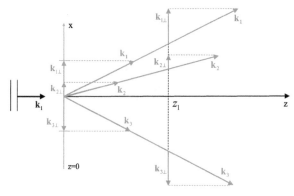

Figure 5.5. Distribution of spatial power spectrum is invariant with z, $|U(\mathbf{k}_\perp,|, z,|, \omega)|^2 \neq f(z)$: the transverse wavevector distribution, e.g., $\mathbf{k}_{1\perp}$, $\mathbf{k}_{2\perp}$, $\mathbf{k}_{3\perp}$ is the same at $z = 0$ and $z = z_1$.

Thanks to the $(\mathbf{k}_\perp, z)$ representation, we only need to perform multiplication operations, which are far faster than the convolution in the spatial domain. However, if we need to express the field in the spatial domain, of course, we end up with the same convolution. Next, we discuss several approximations that make these convolution calculations easier to perform.

5.3 Angular spectrum propagation approximation

In section 5.2 we discovered the interesting features of the $(\mathbf{k}_\perp, z)$ representation when dealing with diffraction problems. In particular, we learned from equation (5.20) that, if a field U is known at a particular plane, say $z = 0$, we can easily find its $\mathbf{k}_\perp$ distribution at an arbitrary plane z, via a simple multiplication (figure 5.6)

$$U(\mathbf{k}_\perp, z; \omega) = i\beta_0 U(\mathbf{k}_\perp, 0; \omega)\frac{e^{i\gamma(\mathbf{k}_\perp)z}}{2\gamma(\mathbf{k}_\perp)}. \tag{5.24}$$

The $U(\mathbf{k}_\perp, z)$ representation of the field is known as the *angular spectrum* because, up to a trivial change of variables, $\mathbf{k}_\perp$ defines the direction of the $\mathbf{k}$–vector, $k_\perp = k \sin(\alpha_3)$, as shown in figure 5.6. For example, $\mathbf{k}$ can be expressed in terms of the direction cosines, i.e., the cosine of the angles $\alpha_1, \alpha_2, \alpha_3$, that it makes with the axes (see section 3.10 in reference [9] and figure 5.6)

$$\mathbf{k} = \beta_0(\cos \alpha_1, \quad \cos \alpha_2, \quad \cos \alpha_3), \tag{5.25}$$

where we note that the cosines are related by

$$\cos \alpha_3 = \sqrt{1 - (\cos \alpha_1)^2 - (\cos \alpha_2)^2}. \tag{5.26}$$

Equation (5.26) is merely another expression of the *dispersion relation* for free space, $|\mathbf{k}| = \beta_0$. While many books use the direction cosines as variables of the angular spectrum, we maintain the wavevector representation, which is more physical, as it relates directly to the wave equation and Fourier analysis.

The propagation of the angular spectrum expressed in equation (5.24) can be simplified further using the following approximation for the amplitude term

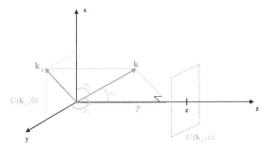

Figure 5.6. Propagation of the angular spectrum from an input plane $z = 0$ to an arbitrary plane, z. One arbitrary wavevector $\mathbf{k}$ is shown. The angular spectrum of the field is the distribution of $\mathbf{k}_\perp$ at each plane z.

$$\frac{1}{\gamma(\mathbf{k}_\perp)} = \frac{1}{\sqrt{\beta_0^2 - k_\perp^2}}$$

$$\simeq \frac{1}{\beta_0} \, .$$

(5.27)

Neglecting $k_\perp$ in $1/\gamma$ is essentially a small angle approximation, as $\mathbf{k}_\perp = \beta_0 \sin \alpha_3$ and $\gamma = \beta_0 \cos \alpha_3$. However, $1/\cos(x)$ is a slowly varying function, which makes this approximation a particularly accurate one. For example, for an angle as large as $30°$, γ is only 15% larger than for $0°$. Combining equations (5.27) and (5.24), we obtain the *angular spectrum propagation approximation*

$$U(\mathbf{k}_\perp, z; \omega) = U(\mathbf{k}_\perp, 0; \omega)e^{i\gamma(\mathbf{k}_\perp)z}, \qquad (5.28)$$

where we ignored the trivial prefactor $i/2$. This approximation removes the $\mathbf{k}_\perp$-dependent amplitude of each plane wave $\exp[i\gamma(\mathbf{k}_\perp)z]$, such that their summation no longer yields the spherical wave. At the same time, angular spectrum propagation keeps the phase term intact, which is far more sensitive than the amplitude. In the next two sections, we discuss approximations in the phase term as well to allow for simplifications in the spatial domain calculations. It has been shown that the angular spectrum propagation is identical to the first Rayleigh–Sommerfeld solution [9, 11].

5.4 Fresnel approximation

In section 5.1, we found that knowing a field at a plane, say $z = 0$, we can predict the diffracted field at a plane z, in the spatial domain, by a 2D convolution with the spherical wave (equation (5.8))

$$U(\mathbf{r}_\perp, z, \omega) = \beta_0 U(\mathbf{r}_\perp, 0; \omega) \circledast_{\mathbf{r}_\perp} \frac{e^{i\beta_0 r}}{r}$$

$$= \beta_0 \int U(\mathbf{r}_\perp', 0; \omega) \frac{e^{i\beta_0 |\mathbf{r}_\perp - \mathbf{r}_\perp', z|}}{|\mathbf{r}_\perp - \mathbf{r}_\perp'|} d^2\mathbf{r}_\perp', \qquad (5.29)$$

where $|\mathbf{r} - \mathbf{r}', z| = \sqrt{(x - x')^2 + (x - y')^2 + z^2}$. Next, we describe some useful approximations of the spherical wave, which makes the integral in equation (5.29) more tractable. When the propagation distance along the z-axis (figure 5.7) is far greater than along the other two axes, the spherical wave can be first approximated by

$$\frac{e^{i\beta_0 r}}{r} \simeq \frac{e^{i\beta_0 r}}{z}. \qquad (5.30)$$

In equation (5.30) we used that the amplitude attenuation is a slow function of x and y, when $z^2 > x^2 + y^2$, such that $1/\sqrt{x^2 + y^2 + z^2} \simeq 1/z$. However, the phase term is significantly more sensitive to x and y variations, such that the next order approximation is needed. Expanding the radial distance in Taylor series, we obtain

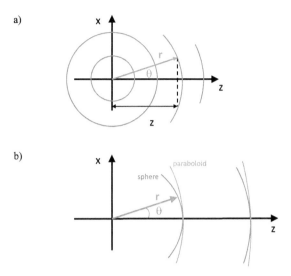

a)

b)

Figure 5.7. Fresnel approximation. For small diffraction angles θ: (a) $/z \cong 1/r$; (b) a paraboloid wavefront becomes a good approximation for the spherical wavefront.

$$r = \sqrt{x^2 + y^2 + z^2}$$

$$= z\sqrt{1 + \frac{x^2 + y^2}{2}} \tag{5.31}$$

$$\simeq z\left[1 + \frac{x^2 + y^2}{2z^2}\right].$$

As a result, the spherical wavelet is now approximated by

$$g(\mathbf{r}, \omega) \simeq \frac{e^{i\beta_0 z}}{z} e^{i\beta_0 \frac{x^2+y^2}{2z}}. \tag{5.32}$$

Equation (5.32) represents the *Fresnel approximation* of the spherical wave. The range of distance z where this approximation holds is called the *Fresnel zone*.

Note that the transverse (x–y) dependence comes in the form of a quadratic (parabolic) phase term. In essence, at plane z, we approximated the spherical wavefront with a paraboloid of revolution (figure 5.7(b)). Notice that, as the distance z increases, the difference between the two wavefronts diminishes. For a given planar (x–y) field distribution at plane $z = 0$, $U(x, y, 0)$, we can calculate the resulting diffracted field at distance z, namely, $U(x, y, z)$ by performing the 2D convolution with the Fresnel wavelet

$$U(\mathbf{r}_\perp, z) = \beta_0 \frac{e^{i\beta_0 z}}{z} U(\mathbf{r}_\perp, 0) \otimes_{\mathbf{r}_\perp} e^{i\frac{\beta_0 r_\perp^2}{2z}}$$

$$= \beta_0 \frac{e^{i\beta_0 z}}{z} \iint U(x', y', 0) e^{i\frac{\beta_0[(x-x')^2+(y-y')^2]}{2z}} dx' dy' \tag{5.33}$$

Equation (5.33) represents the so-called *Fresnel diffraction equation*, which essentially explains the field propagation using Huygens' concept of secondary point sources, except that now each of those secondary point sources emits Fresnel wavelets (parabolic wavefronts) rather than spherical waves. This is the 'traditional', i.e., spatial domain expression for the Fresnel approximation. On the other hand, calculations in the **k**-vector space are again expected to be more direct. Thus, taking the Fourier transform of equation (5.33) with respect to variables (x, y), we obtain a simple product

$$U(\mathbf{k}_\perp, z) = \frac{e^{i\beta_0 z}}{z} U(\mathbf{k}_\perp, 0)e^{-\frac{i}{2\beta_0}\frac{zk_\perp^2}{2\beta_0}}. \tag{5.34}$$

In equation (5.34) we used the Fourier transform property of 2D Gaussian functions, $\exp\left[i\left(\frac{x^2 + y^2}{2b^2}\right)\right] \leftrightarrow \exp\left[-ib^2\left(\frac{k_x^2 + k_y^2}{2}\right)\right]/b$. Note that the prefactor $e^{i\beta_0 z}/z$ is typically neglected, as it contains no information about the x–y field dependence. Typical measurements involve a detector at a fixed plane z, which reports on the x–y distribution of the diffracted field. Such measurements are central to important biophotonics applications such as imaging.

5.5 Fraunhofer approximation

A major simplification in calculating the diffracted field occurs when the transverse extent of the input field is much smaller than the distance z, when the phase term in equation (5.33) satisfies $\beta_0(x'^2 + y'^2) \ll z$ (figure 5.8). Another, intuitive way to express this constraint is

$$\frac{x'}{z} \ll \frac{\lambda}{x'},$$

where λ is the wavelength of light, $\lambda = 2\pi/\beta_0$. In figure 5.8, we see that the maximum extent of the input field subtends an angle $\theta \simeq x'/z$. Thus, the *Fraunhofer* condition requires this angle θ to be much smaller than λ/x''.

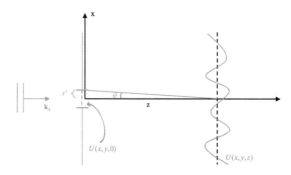

Figure 5.8. Diffraction under the Fraunhoffer approximation.

Under these conditions, the phase term under the convolution of equation (5.33), can be approximated as

$$\frac{\beta_0}{2z}[(x - x') + (y - y')^2] \simeq \frac{\beta_0}{2z}[(x^2 + y^2) - 2(xx' + yy')]$$

(5.35)

$$= \frac{\beta_0}{2z}(x^2 + y^2) - \frac{\beta_0}{z}(xx' + yy').$$

Thus, the convolution integral in equation (5.33) simplifies to

$$U(x, y, z; \omega) = \beta_0 \frac{e^{i\beta_0 z}}{z} e^{i\frac{\beta_0(x^2+y^2)}{2z}} \int U(x', y', 0)e^{-i\frac{\beta_0}{z}(xx'+yy')}dx'dy'.$$

(5.36)

Aside from non-essential prefactors, the integral in equation (5.36) is a 2D Fourier transform. Let us first discuss the relevance of the prefactors. First, $\beta_0 = 2\pi/\lambda$ provides a wavelength-dependent amplitude which becomes important when dealing with broadband fields. As discussed already, $1/z$ gives an attenuation factor with distance, which is important when comparing field amplitudes at different z-planes. The quadratic phase term $\exp[i\beta_0(x^2 + y^2)/2z]$ becomes increasingly relevant as we go further away from the optical axis at the observation plane. When these distances are comparable with those at the input plane, for which the Fraunhofer approximation was made, this quadratic phase term can be neglected as well, $\beta_0(x^2 + y^2) \ll z$. Note that, even when this approximation does not hold, this term vanishes when the amplitude $| U(x, y, z)|$ or irradiance $| U(x, y, z)|^2$ of the diffracted field are of interest. Finally, the plane wave prefactor, $e^{i\beta_0 z}$, simply adds a constant phase shift to U, which is unimportant, given that the origin of the phase delay is arbitrary (only relative phase shifts are of physical importance). Therefore, moving forward, we will keep only the wavelength and z-dependent terms,

$$U(x, y, z; \omega) = \frac{\beta_0}{z}U(k_x, k_y, 0; \omega)\Big|_{\substack{k_x=\frac{\beta_0 x}{z} \\ k_y=\frac{\beta_0 y}{z}}}.$$

(5.37)

In equation (5.37), $U(k_x, k_y, 0)$ is the Fourier transform of $U(x, y, 0)$ evaluated at the spatial frequencies $k_x = \beta_0 x/z$ and $k_y = \beta_0 y/z$, which brings the result in the spatial domain, as the left-hand side.

Equation (5.37) is the main result of the *Fraunhofer, far-zone approximation*. It is a result of highly practical importance, and provides immediate physical insight into the diffraction process. The diffracted field at a distance z is simply the Fourier transform of the input field, evaluated at the frequencies k_x, k_y above. In other words, the diffracted field provides direct access to the frequency content of the input field.

In order to study the physical significance of the spatial frequencies that are directly accessible from measurements of the diffracted field, let us consider figure 5.9. We note right away that $\frac{x}{y}$ and $\frac{y}{x}$ define tangents of an angle. In fact, since z is approximating $\mathbf{r}$, we can conclude that x/z and y/z are the *sines* of two angles, say θ and ϕ (figure 5.9(a)),

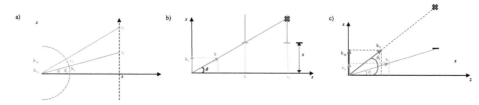

Figure 5.9. (a) Spatial frequencies of the input field generate diffraction angles proportional with these frequencies. (b) Increasing z of the observation plane low-passes the frequencies observed. The **k**-vector passes through the aperture of a detector at plane z_1 but not at plane z_2 (the red cross indicates that the wave is blocked). (c) The same frequency of the input field generates higher diffraction angles for red versus blue light. Therefore, for a fixed observer, or a detector of fixed aperture size, blue light carries higher spatial frequencies.

$$k_x = \beta_0 \sin\theta$$
$$k_y = \beta_0 \sin\phi. \tag{5.38}$$

Thus, k_x and k_y are indeed the projection of the **k**−vector onto the x- and y-axis, respectively. In other words, each spatial frequency at the input plane generates a diffracted plane wave at a particular angle, with the tilt angle proportional to that frequency.

Let us discuss in more detail the meaning of the three variables that occur in the spatial frequency, x (or y), z, and $\beta_0 = 2\pi/\lambda$.

First, if $z = $ const., $\lambda = $ const. (figure 5.9(a))

$$k_x \propto x \tag{5.39a}$$

$$k_y \propto y. \tag{5.39b}$$

This linear dependence indicates that at given plane z and fixed wavelength, accessing the diffracted field at higher coordinates in the observation plane (e.g., $x_2 > x_1$, in figure 5.9(a)) informs about higher frequencies of the input.

Second, $x, y = $ const., $\lambda = $ const. (figure 5.9(b))

$$k_x \propto \frac{1}{z} \tag{5.40a}$$

$$k_y \propto \frac{1}{z}. \tag{5.40b}$$

Equation (5.40) shows that if the wavelength is fixed and the coordinates at the observation plane are fixed, the frequencies accessed decrease with increasing z and vice-versa. For example, moving a camera of a given chip size away from the input plane, filters some high frequencies as high-angle **k**−vectors are blocked.

Third, $x, y = $ const., $z = $ const. (figure 5.9(c))

$$k_x \propto \frac{1}{\lambda} \tag{5.41a}$$

$$k_y \propto \frac{1}{\lambda} \tag{5.41b}$$

According to equation (5.41), at a given observation plane and coordinate, decreasing the wavelength increases the spatial frequencies observed and vice-versa. As shown in figure 5.9(c), assuming the same spatial frequency, $k_x \propto \sin(\theta)/\lambda$, of the input field, the red light will be diffracted at a larger angle than blue light. Thus, for a fixed detector size at plane z, only blue light can be recorded. This simple result has enormous practical implications: for example, in order to measure finer details from an object (higher frequencies) one can decrease the wavelength of investigation. This is the reason why, for accessing the nanoscale, instruments currently use deep-UV radiation, x-rays, and electron waves, with wavelengths much shorter than the visible light. Lithography, which can be regarded as imaging in reverse, follows the same principle. Another example is increasing the storage capacity on optical disks, from near infrared (CD), to red (DVD), and blue light (Blue Ray).

In summary, we can think of the Fraunhofer regime as the situation where different plane waves (propagating at different angles) are generated by different spatial frequencies of the input field. Another way to put this is that, at large distances, the propagation angles corresponding to different spatial frequencies do not mix. This remarkable property allows us to solve many problems of practical interest with extreme ease, by invoking various Fourier transform pairs and their properties, as described in volume 1 [12].

5.6 Fourier properties of lenses

In this section we show that lenses have the capability to perform Fourier transforms, much like free space, and with the added benefit of eliminating the need for a large distance of propagation.

5.6.1 Lens as a phase transformer

Let us consider the biconvex lens in figure 5.10. We would like to determine first the effect that the lens has on an incident plane wave. This effect can be incorporated via a transmission function of the form

$$t(x, y) = e^{i\phi(x, y)}. \tag{5.42}$$

The problem reduces to evaluating the phase delay produced by the lens as a function of the off-axis distance, or the polar coordinate $r_\perp = \sqrt{x^2 + y^2}$ (figure 5.10(b)), namely

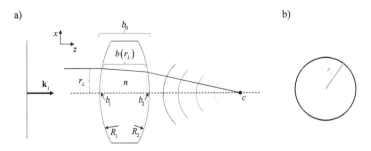

Figure 5.10. (a) Phase transformation by a thin convergent lens. (b) Cross section and polar coordinate r.

$$\phi(r_\perp) = \phi_{glass}(r_\perp) + \phi_{air}(r_\perp)$$
$$= n\beta_0 b(r_\perp) + \beta_0[b_0 - b(r_\perp)] \qquad (5.43)$$
$$= \beta_0 b_0 + (n - 1)\beta_0 b(r_\perp)$$

In equation (5.43), ϕ_{glass} and ϕ_{air} are the phase shifts due to the glass and air portions, respectively, β_0 is the wavenumber in air, $\beta_0 = 2\pi/\lambda$, b_0 is the thickness along the optical axis, i.e., at $r_\perp = 0$ (maximum thickness), $b(r_\perp)$ is the thickness at an off-axis distance $r_\perp$, and n is the refractive index of the glass.

The local thickness, $b(r_\perp)$, can be expressed as

$$b(r) = b_0 - b_1(r_\perp) - b_2(r_\perp), \qquad (5.44)$$

where b_1 and b_2 are the segments shown in figure 5.10, which can be calculated using simple geometry, following the geometry in figure 5.11. For small angles, ABC becomes a right triangle, when the following identity applies (the perpendicular theorem)

$$|AD|^2 = 2|BD| \cdot |DC|. \qquad (5.45)$$

Since $|AD| = r_\perp$, $|BD| = b_1$, and $|DC| = R_1$, we obtain

$$b_1(r_\perp) = \frac{r_\perp^2}{2R_1}. \qquad (5.46)$$

It follows that the thickness, $b(r_\perp)$, can be expressed by combining equations (5.44) and (5.46),

$$b(r_\perp) = b_0 - \frac{r_\perp^2}{2}\left(\frac{1}{R_1} - \frac{1}{R_2}\right). \qquad (5.47)$$

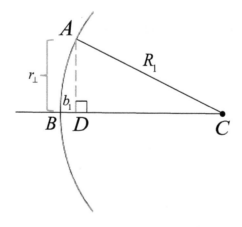

Figure 5.11. Geometry of the small angle propagation through the lens.

Therefore, the phase distribution in equation (5.43) becomes

$$\phi(r_\perp) = \phi_0 - \frac{\beta_0 r_\perp^2}{2}(n-1)\left(\frac{1}{R_1} - \frac{1}{R_2}\right),$$ (5.48)

where $\phi_0 = \beta_0 b_0$.

In equations (5.46)–(5.48), we used the geometrical optics convention whereby surfaces with centers to the left (right) are considered to have negative (positive) radius; in our case $R_1 > 0$ and $R_2 < 0$. We recognize that the focal distance associated with a thin lens is given by the lens maker equation, namely,

$$\frac{1}{f} = (n-1)\left(\frac{1}{R_1} - \frac{1}{R_2}\right).$$ (5.49)

As a result the phase distribution after the lens has a particularly simple expression,

$$\phi(r_\perp) = \phi_0 - \frac{\beta_0 r_\perp^2}{2f}.$$ (5.50)

Finally, the lens transmission function, which establishes how the plane wave incident is transformed by the lens, has the form

$$t(r) = e^{iknb_0}e^{-i\frac{\beta_0 r^2}{2f}}.$$ (5.51)

Equation (5.51) shows that the effect of the lens is to transform a plane wave into a parabolic wavefront. The negative sign $(-i\beta_0 r/2f)$ conventionally denotes a *convergent* field, while the positive sign marks a divergent field. By comparing equations (5.51) and (5.32) it is clear that the effect of propagation through free space is qualitatively similar to transmission through a thin *divergent* (negative f) lens, as illustrated in figure 5.12. This close similarity can be explained by noticing that the negative lens makes the field look as if it originates in a *virtual focal point* at the left of the lens, much like light propagating in free space from a point source.

5.6.2 Lens as a Fourier transformer

Next, let us consider the field propagation through a combination of free space and convergent lens (figure 5.13). The goal is to derive an expression for the output field $U_4(x, y)$ as a function of input field $U_1(x, y)$. This problem can be broken down into a Fresnel propagation over distance d_1, followed by a transformation by the lens of focal distance f, and, finally, a propagation over distance d_2.

In section 5.4 (equations (5.32) and (5.33)) we found that the Fresnel propagation can be described as a convolution with the quadratic phase function (Fresnel wavelet). Thus, the propagation can be written symbolically as

$$U_2(x, y) = U_1(x, y)\circledcirc e^{i\frac{\beta_0(x^2+y^2)}{2d_1}}$$
$$U_3(x, y) = U_2(x, y)e^{-i\frac{\beta_0(x^2+y^2)}{2f}}$$
$$U_4(x, y) = U_3(x, y)\circledcirc e^{i\frac{\beta_0(x^2+y^2)}{2d_2}}$$ (5.52)

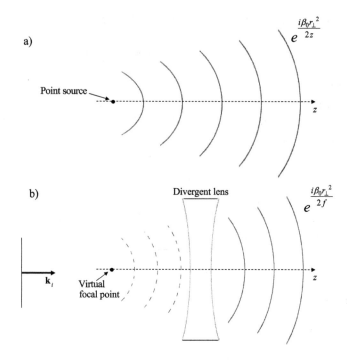

Figure 5.12. (a) Field propagation in free space emitted by a point source. (b) Plane wave propagation through a negative lens. The divergent light appears to be originating from a point to the left of the lens, which is called virtual focal point.

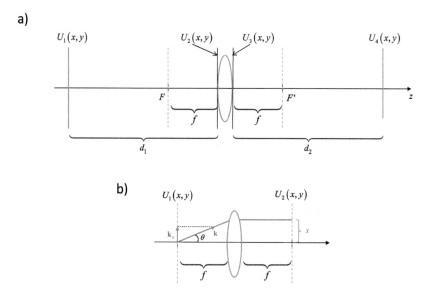

Figure 5.13. (a) Propagation through free space and a convergent lens. The fields at the focal planes F and F' are Fourier transforms of each other. (b) The lens maps each spatial frequency from the input field into a point on the x-axis at the output plane.

Carrying out these calculations is straightforward but somewhat tedious. However, a great simplification arises in the special case where

$$d_1 = d_2 = f. \tag{5.53}$$

Thus, if the input field is at the front focal plane of the lens and the output field is observed at the back focal plane, then the two fields are related via an exact Fourier transform

$$U_4(x, y) = \int\int U_1(x, y)e^{-i(k_{x4}x+k_{y4}y)}dxdy$$
$$k_{x4} = \beta_0\frac{x_4}{f}$$
$$k_{y4} = \beta_0\frac{y_4}{f}, , \tag{5.54}$$

where we ignored trivial factors preceding the integral in equation (5.54). This result is significant, as it establishes a simple, yet powerful way to compute analog Fourier transforms, virtually instantaneously.

It is important to note the similarity of the spatial frequency between the Fraunhofer and lens Fourier transformation of the input field: the only difference is that now z is replaced by f. Therefore, we can reiterate our observations from the Fraunhofer diffraction: shorter wavelengths and large x-coordinates at the output plane will carry higher spatial frequencies from the input field. Furthermore, if these two parameters are fixed, a shorter focal distance will help capture higher frequencies. Generally, high-resolution optical microscopes use objective lenses with short focal distances. As shown in figure 5.13(b), the spatial frequency can be written in terms of the angle emerging for the input field as well, $k_x = \beta_0 \sin(\theta)$. This angle incorporates both the focal distance and the x-coordinate at the output plane. In other words, for a fixed wavelength, $\beta_0 = const.$, $\sin(\theta)$ is the only parameter that decides the maximum spatial frequency captured by the lens from the input field. The *sine* of the maximum angle θ that a lens can transmit is referred to as the *numerical aperture (NA)* of the lens, $NA = \sin(\theta_{max})$. Thus, the highest spatial frequency captured by the lens is $k_x^{max} = \beta_0 NA$, which governs the maximum resolution with which the lens can resolve the input field. For this reason, the NA is one of the critical parameters for microscope objectives and camera lenses.

5.7 Problems

1. A plane wave, $e^{i\beta_0 z}$, is incident on a circular aperture of radius a (figure 5.14).
   a) Write an expression for the field at the aperture using our usual elementary functions, e.g., the rectangular function.
   b) Calculate the field U at a plane z behind the aperture in the angular spectrum representation ($\mathbf{k}_\perp, z$), without angular approximations.
   c) Express the field $U(\mathbf{k}_\perp, z)$ under the angular spectrum propagation approximation.

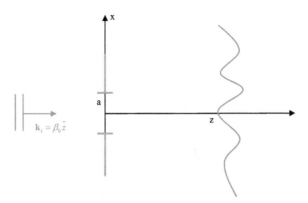

Figure 5.14. Problem 1.

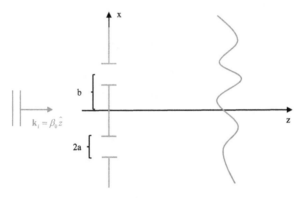

Figure 5.15. Problem 5.

d) Calculate the field at the plane z under the Fresnel approximation, in the $(\mathbf{k}_\perp, z)$ representation.

e) Calculate the field at plane z under the Fraunhofer approximation, in the $(\mathbf{k}_\perp, z)$ domain.

f) Express the field in (b), (c), (d), (e) in the spatial domain (x, y, z).

g) Consider $\lambda = 1\ \mu m$, $a = 10\ \mu m$. Plot the profile $|U^2(x)|$ at $z = 1$ cm, $z = 1\ m$, $z = 100$ m, for the four expressions obtained in (e). For each z-value, plot the four profiles on the same graph and discuss the accuracy of each approximation.

2. Redo problem 1 for a square aperture of side a.

3. Redo problem 1 if the incident plane wave makes an angle $\theta = 30°$ with the z-axis.

4. Redo problem 1 if the incident plane wave makes an angle $\theta = 30°$ with the z-axis and the aperture is a square of side a.

5. Calculate the diffracted field $U(\mathbf{k}_\perp, z)$ and $U(x, y, z)$ generated by a double slit under *on*-axis, plane wave illumination, as shown in figure 5.5. Express the field first with no angular approximations and then under the angular spectrum propagation, Fresnel, and Fraunhofer approximation (figure 5.15).

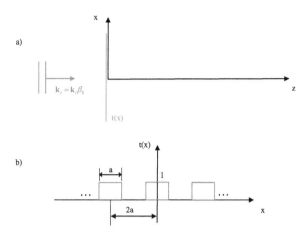

Figure 5.16. Problem 10. (a) Plane wave incident on an amplitude grating of transmission function t. (b) Transmission function of the grating.

6. Redo problem 5 for an arbitrary incident plane wave, $\mathbf{k}_i = \beta_0 \hat{\mathbf{k}}_i$.
7. Redo problem 5 for two square apertures of side a.
8. Redo problem 5 for two circular apertures of radius a.
9. Redo problems 7 and 8 for an arbitrary incident plane wave, $\mathbf{k}_i = \beta_0 \hat{\mathbf{k}}_i$.
10. Under the Fraunhofer approximation, calculate the diffracted field generated by a plane wave incident on an infinite amplitude transmission grating of period $2a$ and square groove width a.
11. Redo problem 10, if the grating has a finite size $b > a$.
12. Redo problem 10 in the case of a *phase* grating, of transmission function $e^{i\phi(x)}$ of the same profile as in figure 5.16(b), $\phi(x) = t(x)$.
13. Redo problem 10 for a sinusoidal *amplitude* grating of period a.
14. Redo problem 13 for a sinusoidal *phase* grating of period a.
15. An infinite 2D array consists of circular apertures of radius a distributed with a period b (figure 5.17). Calculate the diffracted field under the Fraunhofer approximation.
16. Redo problem 15 for square apertures of side a.
17. Redo problem 15 for a finite array cut into a circular shape of radius $c > b$.
18. Redo problem 16 for a finite array cut into a square shape of side $c > b$.
19. Write the expression for the field $U(\mathbf{k}_\perp, z)$ diffracted by a screen of transmission $t(x, y)$ when illuminated by two plane waves that are symmetric with respect to the z-axis (figure 5.18). First write the exact solution in the $(\mathbf{k}_\perp, z)$ representation, then use the angular spectrum propagation approximation, the Fresnel and, finally, Fraunhofer approximation.
20. Calculate the exact expression for the diffracted field in the $(\mathbf{k}_\perp, z)$ domain diffracted by a transparent screen illuminated by a spherical wave, as shown in figure 5.19.
21. Redo problem 20, when the point source is off-axis (figure 5.20).

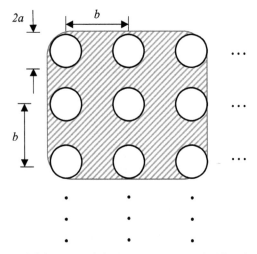

Figure 5.17. Problem 15: 2D infinite array of circular apertures. Hatched lines indicate the opaque region.

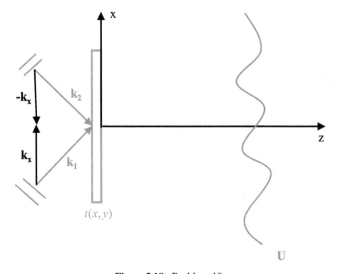

Figure 5.18. Problem 19.

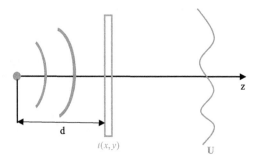

Figure 5.19. Problem 20.

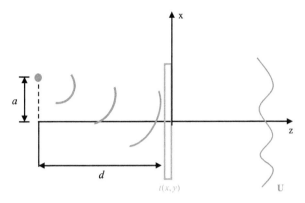

Figure 5.20. Problem 21.

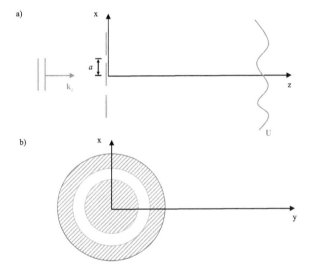

Figure 5.21. Problem 22. (a) On-axis plane wave incident on a ring-shaped aperture. (b) Ring aperture in the x–y (transverse) plane.

22. Under the Fraunhofer approximation, calculate the diffraction by a ring-shaped aperture when illuminated by an on-axis plane wave (figure 5.21). Assume an infinitely thin ring.

23. Redo problem 21, when the thickness of the ring is b.

24. The light diffracted by a screen $t(x, y)$ is transformed by a *telecentric* lens system (the back face plane of lens 1 coincides with the front faced plane of lens 2), as shown in figure 5.22. Calculate the field at the back focal plane of lens 2, under the Fresnel approximation.

25. Calculate the field at the back focal plane of the lens, when the transparency is placed right in front of the lens (figure 5.23). Use the Fresnel approximation.

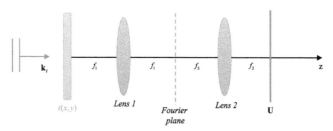

Figure 5.22. Problem 24.

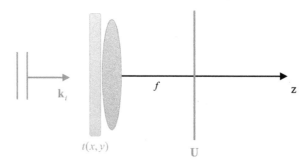

Figure 5.23. Problem 25.

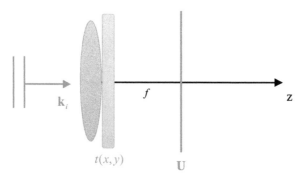

Figure 5.24. Problem 26.

26. Redo problem 25, when the transparency is right behind the lens (figure 5.24).
27. Redo problem 24, when the transparency is placed at a distance d behind the lens (figure 5.25).
28. A transparency is placed in the front focal plane of a lens and is illuminated by a spherical wave, as shown in figure 5.26. Use the Fresnel approximation to calculate
    a) The field at the back focal plane of the lens.
    b) The plane z_0 at which the field approaches the Fourier transform of the transmission function.

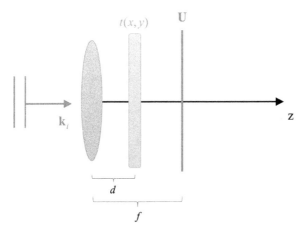

Figure 5.25. Problem 27.

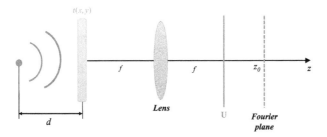

Figure 5.26. Problem 28.

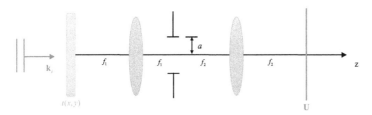

Figure 5.27. Problem 29.

29. Consider the geometry in figure 5.27, which is similar to that in figure 5.22, except, now we have a circular aperture placed at the focal plane. Calculate the field at the back focal plane of lens 2, under the Fresnel approximations.

30. Redo problem 29 when the incident field is an off-axis plane wave. Compare the result with that in problem 29.

References

[1] Sommerfeld A 1950 *Lectures on Theoretical Physics: Optics* (New York: Academic) p v

[2] Popescu G *Principles of Biophotonics, Volume 6—Light Propagation in Inhomogeneous Media* (Bristol: IOP Publishing) (not yet published)

[3] Born M and Wolf E 2013 *Principles of Optics: Electromagnetic Theory of Propagation, Interference and Diffraction of Light.* (Amsterdam: Elsevier)

[4] Newton I 1718 *Opticks.* 2nd edn. (London: W. and J. Innys) 4 1

[5] Huygens C 1912 *Treatise on Light: In Which Are Explained the Causes of that Which Occurs in Reflexion, & in Refraction. And Particularly in the Strange Refraction of Iceland Crystal.* (London: MacMillan and Company Limited)

[6] Young T 1807 *A Course of Lectures on Natural Philosophy and the Mechanical Arts* vol 1 (London: J. Johnson)

[7] Fresnel A 1818 *Mémoire sur la diffraction de la lumière.* (Paris: Mémoires de l'Académie des Sciences) V 339–475

[8] Crew H 1900 *The Wave Theory of Light.* vol 10 (New York: American Book Company)

[9] Goodman J W 1996 *Introduction to Fourier Optics.* 2nd edn (Series in Electrical and Computer Engineering) (New York: McGraw-Hill) pp. xviii, 441

[10] Weyl H 1919 *Ann. Physik.* **60** 481–500

[11] Sherman G C 1967 Application of the convolution theorem to Rayleigh's integral formulas *JOSA* **57** 546–47

[12] Popescu G 2018 *Principles of Biophotonics, Volume 1—Linear Systems and the Fourier Transform in Optics* (Bristol: IOP Publishing)

IOP Publishing

Principles of Biophotonics, Volume 3
Field propagation in linear, homogeneous, dispersionless, isotropic media
Gabriel Popescu

Chapter 6

Geometrical optics

6.1 Applicability of geometrical optics

Geometrical or *ray* optics is the simplest theory of light propagation. Nevertheless, it well approximates common experimental situations, such as image formation in an optical system. The main concept in geometrical optics is that in homogeneous media light travels along straight lines, i.e., *rays*. Thus, predicting how rays propagate, e.g., where they cross each other, comes down to solving geometry problems (hence, the name of the theory). Ray optics is limited in describing wave phenomena, such as diffraction (discussed in chapter 5) and only applies when such effects are negligible. The general rule for its applicability is that the objects with which the light interacts have spatial features much larger than the wavelength, λ. Equivalently, we can say that geometrical optics applies in the limit $\lambda \to 0$.

Another way to look at this approximation is to connect it with Fraunhofer diffraction (section 5.5) and reiterate that low spatial frequencies generate low diffraction angles. Thus, geometrical optics is also a small-angle approximation. The rays are considered to travel almost parallel to the optical axis, hence the term *paraxial approximation* that is sometimes used synonymously.

In this chapter, we discuss the main principles of geometrical optics and establish a matrix formalism that is extremely practical for ray propagation.

6.2 WKB approximation: eikonal equation and geometrical optics

The Wentzel–Kramers–Brillouin (WKB) approximation applies to slowly varying (*smooth*) inhomogeneous media, in which we can assume that the amplitude of the field is approximately constant and the phase does not change abruptly. This treatment results in the *geometrical optics* approximation of the wave equation, which is referred to as the *eikonal equation*.

We start by writing the scalar field explicitly in terms of an amplitude and phase, $U(\mathbf{r}, \omega) = A(\mathbf{r}, \omega)e^{i\phi(\mathbf{r}, \omega)}$, such that the Laplacian in the Helmholtz equation reads

$$\nabla^2[A(\mathbf{r}, \omega)e^{i\phi(\mathbf{r}, \omega)}]$$
$$= \nabla \cdot [e^{i\phi(\mathbf{r}, \omega)}\nabla A(\mathbf{r}, \omega) + e^{i\phi(\mathbf{r}, \omega)}A(\mathbf{r}, \omega)i\nabla\phi(\mathbf{r}, \omega)]. \tag{6.1}$$

The WKB approximation states that the amplitude is spatially constant and that $\phi(\mathbf{r})$ fluctuates slowly in space, specifically, the following approximations apply

$$A(\mathbf{r}, \omega) \simeq \text{const.}(\mathbf{r}) \tag{6.2a}$$

$$\nabla A(\mathbf{r}, \omega) \simeq 0 \tag{6.2b}$$

$$\nabla^2 A(\mathbf{r}, \omega) \simeq 0 \tag{6.2c}$$

$$\nabla^2 \phi(\mathbf{r}, \omega) \simeq 0 \tag{6.2d}$$

With these approximations, equation (6.1) simplifies to

$$\nabla^2 U(\mathbf{r}, \omega) \simeq -U(\mathbf{r}, \omega) \mid \nabla\phi(\mathbf{r}, \omega) \mid^2 + iU(\mathbf{r}, \omega)\nabla^2\phi(\mathbf{r}, \omega),$$
$$\simeq -U(\mathbf{r}, \omega) \mid \nabla\phi(\mathbf{r}, \omega) \mid^2, \tag{6.3}$$

where $\mid \nabla\phi(\mathbf{r}, \omega) \mid^2$ is the *gradient intensity* of ϕ, defined as

$$\mid \nabla\phi(\mathbf{r}, \omega) \mid^2 = \left(\frac{\partial\phi(\mathbf{r}, \omega)}{\partial x}\right)^2 + \left(\frac{\partial\phi(\mathbf{r}, \omega)}{\partial y}\right)^2 + \left(\frac{\partial\phi(\mathbf{r}, \omega)}{\partial z}\right)^2. \tag{6.4}$$

A more accurate approximation, the *Rytov approximation*, retains the $\nabla^2\phi$ contribution (see volume 6 [1] on light scattering).

Inserting the expression for $\nabla^2 U$ into the Helmholtz equation (section 5.1), we have

$$\{\beta^2(\mathbf{r}, \omega) - [\nabla\phi(\mathbf{r}, \omega)]^2\}U(\mathbf{r}, \omega) = 0,$$
$$\beta(\mathbf{r}, \omega) = n(\mathbf{r}, \omega)\beta_0 \, . \tag{6.5}$$

If we introduce the *optical path length* associated with the field as

$$s(\mathbf{r}, \omega) = \phi(\mathbf{r}, \omega)/\beta_0, \tag{6.6}$$

we can express equation (6.5) in terms of the refractive index, $n = \beta/\beta_0$, and s,

$$[\nabla s(\mathbf{r}, \omega)]^2 = n^2(\mathbf{r}, \omega). \tag{6.7}$$

Equation (6.7) is known as the *eikonal* (Greek for 'image') *equation* and is central to geometrical optics. Recall (see appendix B in volume 1 [2]) that the infinitesimal change in path length is

$$ds(\mathbf{r}, \omega) = \nabla s(\mathbf{r}, \omega) \cdot d\mathbf{r}. \tag{6.8}$$

The surface of constant pathlength, $s(\mathbf{r}) = \text{const.}$, is, by definition, the wavefront associated with the optical field. Equation (6.8) indicates that pathlength change, ds, only occurs along a direction parallel to the gradient. This is illustrated in figure 6.1(a). It follows that any ray crosses the wavefront on a direction perpendicular to it (figure 6.1(b)). Projecting $d\mathbf{r}$ ($d\mathbf{r}\perp\mathbf{r}$) in a direction parallel to $\nabla s(\mathbf{r})$, $d\mathbf{r}_{\parallel} = d\mathbf{l}$, and one perpendicular, $d\mathbf{r}_{\perp}$, we see that

a)   b)   c)

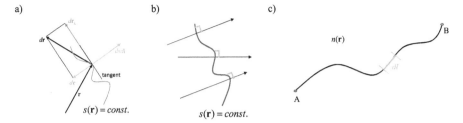

Figure 6.1. (a) The pathlength accumulates on a direction normal to the wavefront, defined as $s(r)$ = const. (b) Rays always cross the wavefront surface along perpendicular directions. (c) The pathlength between two points is a line integral.

$$ds(\mathbf{r}, \omega) = \nabla s(\mathbf{r}, \omega)dl\,\hat{d\mathbf{l}}$$
$$= n(\mathbf{r}, \omega)dl. \tag{6.9}$$

Therefore, the pathlength can be expressed as the integral of the refractive index over the path, l,

$$S_{AB}(\mathbf{r}, \omega) = \int_{A}^{B} n(\mathbf{r}, \omega)dl$$
$$= \int_{0}^{L} n(\mathbf{r}, \omega)dl, \tag{6.10}$$

where L is the total length of the ray that connects A and B. This path integral is illustrated in figure 6.1(c).

The WKB approximation and the eikonal equation are very useful for describing light propagation in transparent, smooth objects, such as optical components and even thin biological specimens. The assumption of negligible amplitude modulation and phase Laplacian is very strong and adds constraints on the smallest features (highest spatial frequencies) that can be suitably described by the WKB approximation. The concept of ray propagation and pathlength defined as the integral in equation (6.10) are commonly used in imaging systems.

6.3 Fermat's principle

In the previous section we derived the *eikonal equation* that governs geometrical optics and provides a connection to the wave equation. However, the ray-like propagation of light was postulated centuries before the discovery of Maxwell's equation. Mo Zi, a Chinese philosopher, politician, and scientist, considered that light propagates in straight lines, as early as the fifth century B.C.E. (for a translation, see [3]). In 300 B.C.E., Euclid wrote *Optica*, which laid out the principles of ray optics. In 214–212, Archimedes is credited with describing the 'Siege of Syracuse', where 'heat rays' were used as a weapon. In 40 C.E., Heron of Alexandria published the work *Catoptrica* in which he postulates that 'the path taken by a ray of light reflected from a plane mirror is shorter than any other reflected path'. Heron's principle is a particular case of that proposed by Fermat in the seventeenth century (see [4] for a biography of Pierre de Fermat).

In its simplest form, Fermat's principle states that light travels between two points on a path of shortest time. Of course, in a homogeneous medium of constant refractive index, this path is a straight line (figure 6.2(a)). The time it takes to connect two points in this case is simply

$$t_{AB} = \frac{L_{AB}}{v}$$
$$= \frac{nL_{AB}}{c}, \tag{6.11}$$

where t_{AB} is the time, L_{AB} the distance, n the constant refractive index, and c the speed of light in a vacuum. We notice that nL_{AB} is the optical pathlength, s, introduced in section 6.2.

Remarkably, this principle holds for inhomogeneous media as well, where the light path is curved (figure 6.2(b)). In this case, the time of travel between A and B is (see figure 6.3(a))

$$t_{AB} = \frac{s}{c}$$
$$= \frac{1}{c} \int_A^B n(\mathbf{r})dl. \tag{6.12}$$

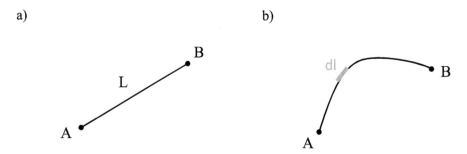

Figure 6.2. Fermat's principle: light travels on a path of minimum time. (a) Homogenous media; (b) inhomogeneous media.

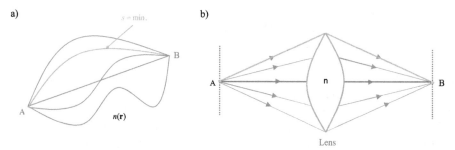

Figure 6.3. (a) The points A and B are connected by light via the shortest optical pathlength (or time). (b) Point B is the image of A through the lens. All rays follow equal pathlengths.

There are particular situations when multiple paths have the same travel time. For example, imaging a point source through a lens, creates such a situation where all the rays connecting the object and image points travel equal pathlengths (figure 6.3(b)). Rays traveling at low angles with respec (green in figure 6.3(b)) encounter longer paths in air and thinner portions of the glass. An ideal lens perfectly compensates the air and glass paths, such that the optical pathlength, or travel time, is the same for all rays. Remarkably, this property holds for all the points in the object plane and all corresponding points in the images plane. As a result, under perfect imaging conditions (no aberrations), the lens is relaying to the image plane a perfect replica of both the amplitude and phase of the object, up to a scaling factor (the transverse magnification of the image). This property plays an important role in imaging, especially when using coherent illumination.

Thus, to include this particular situation, we can state Fermat's principle by requiring the optical pathlength to be stationary,

$$\delta s = \delta \int_A^B n(\mathbf{r}) dl \qquad (6.13)$$
$$= 0.$$

If we consider that the two points are in two media of different refractive indices, applying Fermat's principle leads to Snell's law. Recall that we derived this law in section 2.6 from two boundary conditions of Maxwell's equation. However, this result precedes Maxwell's work, as it was discovered first by Ibn Sahl in the tenth century and rediscovered later by Willebrord Snellius in 1621.

It is left as an exercise (problem 3) to prove that, in figure 6.4, the path of minimum time yield's Snell's law,

$$n_1 \sin \theta_1 = n_2 \sin \theta_2. \qquad (6.14)$$

Note that as the ray passes into a higher refractive index, $n_2 > n_1$, it bends toward the normal, $\theta_2 < \theta_1$, and vice-versa. The sign convention extends to distances as well, thus, avoiding ambiguities as to whether a distance is measured to the right or left of a reference point. Similarly, focal distances and surface radii can be positive or negative as well. All sign conventions are summarized in table 6.1 and figure 6.5.

The law of reflection is obtained by replacing n_2 with $-n_1$ in Snell's law, namely

$$\theta_1 = -\theta_2 \qquad (6.15)$$

This sign change accounts for the sign convention, which states that, starting from the normal, the angle is considered positive if it corresponds to a counter clockwise rotation. As already discussed in section 2.6., Snell's law yields a critical angle of incidence, $\theta_c = \sin^{-1}(n_2/n_1)$, for which total internal reflection is achieved (only when $n_2 < n_1$).

6.4 Refraction through curved surfaces

Let us consider the refraction phenomenon at a spherical boundary (figure 6.6(a)). We aim to find a relationship between the input angle θ_1 and output angle θ_2. The surface is characterized by a radius R and center C. Note that Snell's law applies to

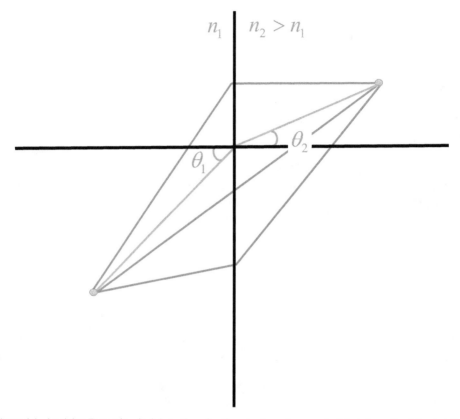

Figure 6.4. Applying Fermat's principle to the refraction at a boundary, we find that the path of shortest time (in green) is the one that satisfies Snell's law.

the angles measured with respect to the normal at the surface, defined as the line connecting the incidence point and the center (figure 6.6(b)),

$$n_1\alpha_1 = n_2\alpha_2 \tag{6.16}$$

Using the geometry in figure 6.6(b), we can find a relationship between the angles measured with respect to the optical axis, θ_1, θ_2 and α_1, α_2,

$$\alpha_1 = \theta_1 + \phi \tag{6.17a}$$

$$\alpha_2 = \theta_2 + \phi \tag{6.17b}$$

$$\phi = \frac{x}{R} \tag{6.17c}$$

where x is the distance between the incidence point and the optical axis. This quantity is called ray *elevation*.

Using the paraxial approximation, $\sin x \simeq x$, Snell's law yields

$$n_1\left(\theta_1 + \frac{x}{R}\right) = n_2\left(\theta_2 + \frac{x}{R}\right), \tag{6.18}$$

Table 6.1. Sign convention

1. The optical axis is the z axis, which is positive to the right of the figure. Rotational symmetry exists around the z axis
2. Distances measured to the left of a reference point are negative, while those measured to the right are positive.
3. Light travels from left to right ($-z$ to $+z$):
   — Left to right → (+) index of refraction,
   — Right to left → (−) index of refraction.
4. Heights are positive in the upward direction.
5. Angles:
   — Measured counterclockwise from a reference are positive;
   — Measured clockwise from a reference are negative.
6. Focal lengths:
   — Converging lens → positive,
   — Diverging lens → negative.
7. Surface radii:
   — Positive means that the center of curvature (C) is to the right of the surface;
   — Negative means that the center of curvature (C) is to the left of the surface.
8. Signs of all indices of refraction are reversed following a reflection.
9. Signs of all distances following a reflection are consistent with our sign convention.

or

$$n_2\theta_2 = n_1\theta_1 - \frac{x}{R}(n_2 - n_1). \tag{6.19}$$

Note that when $R \to \infty$, i.e., the surface becomes flat, we obtain the usual form of Snell's law. Equation (6.19) allows us to obtain the focal distance of this surface, defined as the point at which an incident horizontal ray ($\theta_1 = 0$) crosses the optical axis (figure 6.6(c)). Thus, we see that the focal distance f is (θ_2 is negative)

$$f = \frac{x}{-\theta_2}. \tag{6.20}$$

Using θ_2 from equation (6.19), for $\theta_1 = 0$, we obtain

$$f = \frac{n_2}{(n_2 - n_1)}R. \tag{6.21}$$

Note that the radius is positive for this *convex* surface, i.e., $R > 0$, where the center is to the right of the surface. It is left as an exercise to use Snell's law and prove that $f = -[n_2/(n_2 - n_1)]R$ for a *concave* surface.

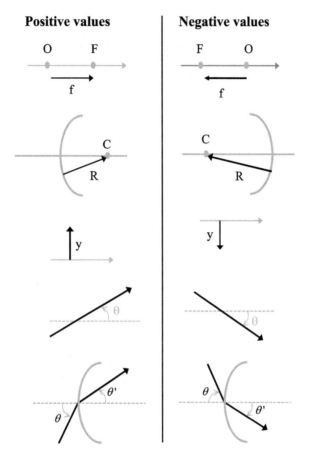

Figure 6.5. Sign convention for distances and angles.

A spherical lens consists of a piece of glass polished to have two spherical surfaces of radii R_1 and R_2 (figure 6.7(a)). Let the refractive index of the lens be n and the surrounding medium air, $n = 1$. In order to find the angle of the ray refracted by the lens, we apply Snell's law twice, for each of the two surfaces (figure 6.7(b)) and ignore the lens thickness, i.e., we assume a thin lens.

At the first interface, the relationship between the angles θ, with respect to the horizontal, are

$$n\theta_2 = -\frac{x}{R_1}(n - 1) + \theta_1 \tag{6.22}$$

Applying the same formula for the second surface, where we see that θ_3 is clockwise, thus, negative,

$$-\theta_3 = -\frac{x}{R_2}(1 - n) + n\theta_2 \tag{6.23}$$

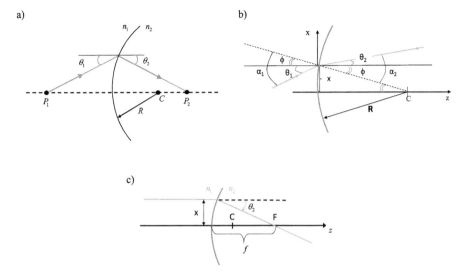

Figure 6.6. (a) Refraction at a spherical surface. (b) Snell's law at a spherical surface. (c) Focal distance of a *convex* surface.

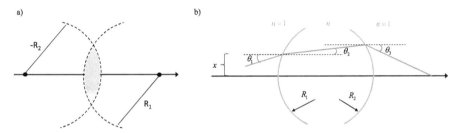

Figure 6.7. (a) Biconvex lens. (b) Refraction through a lens.

Adding equations (6.22) and (6.23) to eliminate θ_2, we obtain

$$-\theta_3 = \theta_1 + \left(\frac{x}{R_1} - \frac{x}{R_2} \right)(n - 1). \tag{6.24}$$

Again, for a flat surface, we obtain $\theta_1 = \theta_2$, as expected. In order to find the focal distance of the lens, we set $\theta_1 = 0$ and determine the distance where the ray intercepts the optical axis,

$$f = \frac{x}{\theta_3}, \tag{6.25}$$

which yields

$$\frac{1}{f} = (n - 1)\left(\frac{1}{R_1} - \frac{1}{R_2} \right). \tag{6.26}$$

This formula, known as the *lens maker equation*, is used very often in practice, as it gives the focal distance of a thin, *spherical* lens. We point out that, if the two surfaces have the same radii, R, maintaining the sign convention for the radii is crucial, namely, $R_1 = R$, $R_2 = -R$, such that the focal distance is

$$\frac{1}{f} = (n - 1)\frac{2}{R}. \tag{6.27}$$

Computing the focal distances for a variety of other lens types is left as an exercise.

6.5 Reflection by curved mirrors

We found earlier that the *reflection law*, $\theta_2 = -\theta_1$, is a simpler version of Snell's law, with no dependence on the refractive index. An important implication of this result is that the wavelength dependence of the refractive index and, thus, of a lens focal distance, disappears if one uses curved mirrors to focus the light. Newton studied color dispersion through prisms and was familiar with chromatic aberrations. His treatise on optics, published in 1718, was entitled *Optics: or a treatise of the Reflections, Refractions, Inflections and Colours of Light* [5]. With this understanding, Newton proposed the *reflection* telescope, in which lenses were replaced by curved mirrors to eliminate dispersion. Next, we study the reflection by curved mirrors of various shapes.

6.5.1 Spherical mirrors

Let us consider the reflection by a ray parallel to the optical axis (Oz) at a concave mirror of radius R (figure 6.8(a)). The goal is to find the point F where the ray intercepts the optical axis and derive an expression for the distance $| OF |$, which plays the role of the *focal distance*. In triangle AFC, we apply the 'sine theorem', namely,

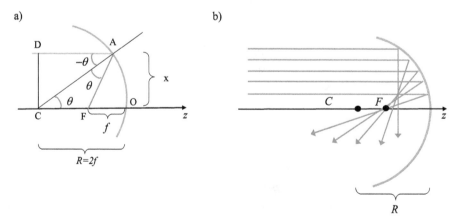

Figure 6.8. (a) Reflection of a horizontal ray by a concave mirror. (b) Elevation-dependent focal distance for large angles of incidence.

$$\frac{\sin(\sphericalangle CAF)}{|CF|} = \frac{\sin(\sphericalangle CFA)}{|AC|}, \tag{6.28}$$

which yields

$$\frac{\sin(\theta)}{R - f} = \frac{\sin(\pi - 2\theta)}{R}. \tag{6.29}$$

Solving for f, we obtain

$$f = R\left(1 - \frac{\sin\theta}{\sin 2\theta}\right)$$

$$= R\left(1 - \frac{1}{2\cos\theta}\right). \tag{6.30}$$

Since $\sin\theta = \dfrac{x}{R}$ (triangle ADC), we see that, generally, the ray intercepts the z-axis at a point that depends on the elevation x, and the focal distance is $f(x)$. In other words, rays at different elevation are focused at different points on the z-axis. This phenomenon is a form of *geometric aberration*, illustrated in figure 6.8(b) and will be discussed in more detail in volume 9 [6].

From equation (6.30), we can calculate the variation of f with θ,

$$\frac{df}{d\theta} = -R\frac{\sin\theta}{2\cos^2\theta}. \tag{6.31}$$
$$\leqslant 0$$

Equation (6.31) indicates that rays of higher elevation are focused closer to the lens apex (O), i.e., produce smaller focal distances, as seen in figure 6.8(b). However, for small angles of incidence, $\cos\theta \simeq 1$, and equation (6.30) simplifies to

$$f = \frac{R}{2} \tag{6.32}$$

We conclude that for horizontal rays of low elevation, or $x/R < <1$, a spherical lens acts as a perfect lens, of focal distance $R/2$.

6.5.2 Parabolic mirrors

It turns out that there exists a surface shape for which a mirror focuses parallel rays to a single point, independent of their elevation with respect to the optical axis. This surface is a *paraboloid of revolution*, described by the equation

$$z = \frac{x^2 + y^2}{b^2}. \tag{6.33}$$

As shown in figure 6.9, all rays converge to a single focal point, which defines a unique focal distance, f. This important property is the reason many telescopes and satellite antennas use parabolic mirrors.

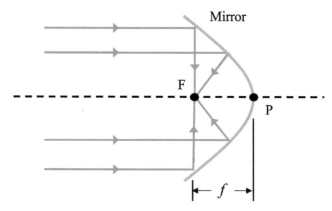

Figure 6.9. Focusing light by a paraboloidal mirror.

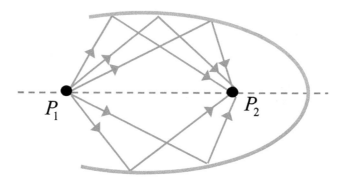

Figure 6.10. Elliptical mirror.

6.5.3 Elliptical mirrors

Another mirror shape of particular importance is the *ellipsoid of revolution* (or *spheroid*), described by

$$\frac{x^2 + y^2}{a^2} + \frac{z^2}{b^2} = 1. \tag{6.34}$$

Figure 6.10 shows the $x - z$ cross section of such a mirror (an ellipse). The ellipse has the well-known property that the sum of the distances between a point on the surface and the two foci is constant. According to Fermat's principle, rays originating from one of the ellipses focus points converge to the second focus. This is another example of a stationary case for Fermat's principle. This propriety of elliptical mirrors allows them to be used for imaging, relaying a point source (or focused light) to another point, without geometric aberrations.

6.6 Ray propagation (ABCD) matrices

In this section we establish a matrix algebra formalism that can be used to solve most geometrical optics problems more efficiently. We start by recognizing that a ray at a

a)   b)

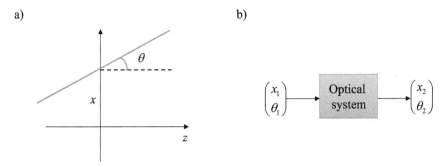

Figure 6.11. (a) At a given plane z, a ray is uniquely identified by a 2D vector $(x, \theta)^{\mathrm{T}}$. (b) An optical setup operating as a linear system in the $(x, \theta)^{\mathrm{T}}$ vector space.

certain plane z is fully characterized by two parameters: its elevation x and its angle with the horizontal (inclination) θ (see figure 6.11(a)).

The goal is to find the operators that characterize various optical systems and provide an input–output relationship in the $(x, \theta)^{T}$ vector space. By optical system we mean any system that is capable of modifying a ray property, such as, free space, lenses, interfaces, and their combinations, etc. The system operator will be in the form of a 2×2 matrix **M**, such that the input–output relationship can be expressed as

$$\begin{pmatrix} x_2 \\ \theta_2 \end{pmatrix} = \begin{pmatrix} A & B \\ C & D \end{pmatrix}\begin{pmatrix} x_1 \\ \theta_1 \end{pmatrix}. \tag{6.35}$$

Next, we will identify this matrix for various optical systems of interest, under the paraxial approximation.

6.6.1 Free space translation

The simplest case of ray propagation is in free space from a plane to another (figure 6.12). We see right away that propagation in free space does not change the angle θ, thus,

$$\theta_2 = \theta_1 \tag{6.36a}$$

From the geometry and small angle approximation, we find the following expression for the elevation x_2

$$x_2 = x_1 + d\theta_1. \tag{6.36b}$$

Equations (6.36a) and (6.36b) can be combined in a vector form as

$$\begin{pmatrix} x_2 \\ \theta_2 \end{pmatrix} = \begin{pmatrix} 1 & d \\ 0 & 1 \end{pmatrix}\begin{pmatrix} x_1 \\ \theta_1 \end{pmatrix}. \tag{6.37}$$

Thus, the ray transfer matrix has a particularly simple form for propagation or translation over a distance d.

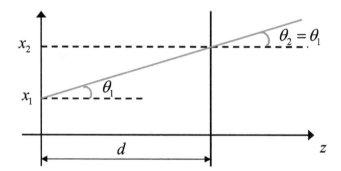

Figure 6.12. Ray propagation in free space.

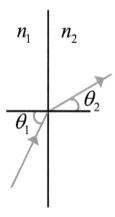

Figure 6.13. Finding the ABCD matrix for refraction at a planar interface.

6.6.2 Refraction through a planar interface

We consider next the ray refraction at a planar interface (figure 6.13). Applying Snell's law for small angles,

$$\theta_2 = \frac{n_1}{n_2}\theta_2. \tag{6.38}$$

Note that we are only describing the ray refraction from 'right' before the interface to 'right' after, thus

$$x_2 = x_1 \tag{6.39}$$

Therefore, the matrix for refraction at a planar interface is

$$M = \begin{pmatrix} 1 & 0 \\ 0 & \frac{n_1}{n_2} \end{pmatrix}. \tag{6.40}$$

6.6.3 Refraction through a spherical interface

Next, let us consider the spherical interface in figure 6.14. Again, $x_2 = x_1$. Using Snell's law for a spherical surface derived earlier (equation (6.19), section 6.4), we have

$$\theta_2 = -\frac{x}{R}\left(\frac{n_2 - n_1}{n_2}\right) + \frac{n_1\theta_1}{n_2} \tag{6.41}$$

Thus, the matrix in this case reads

$$M = \begin{pmatrix} 1 & 0 \\ -\dfrac{n_2 - n_1}{n_2 R} & \dfrac{n_1}{n_2} \end{pmatrix},$$
$$= \begin{pmatrix} 1 & 0 \\ -\dfrac{1}{f} & \dfrac{n_1}{n_2} \end{pmatrix} \tag{6.42}$$

where f is the focal length of the spherical surface derived in section 6.4 (equation (6.20)). When $R \to \infty$, we recover the matrix for the flat surface, as expected. Note that for a *concave* surface, $R < 0$.

6.6.4 Transmission through a thick lens

Since we know the translation and refractive matrices, we can cascade them and create any combination of layers separated by spherical surfaces. In particular, we can study the propagation through a lens, by considering a translation sandwiched by two refractions. Consider the thick lens in figure 6.15, of refractive index n and thickness t. Recall that, when discussing refraction through a lens in section 6.4, we ignored its thickness. The input–output relationship for the ray vector is

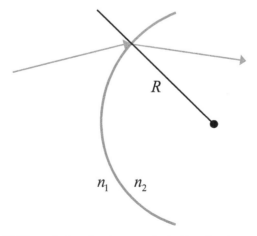

Figure 6.14. Finding the ABCD matrix for refraction at a spherical interface (convex: $R > 0$; concave: $R < 0$).

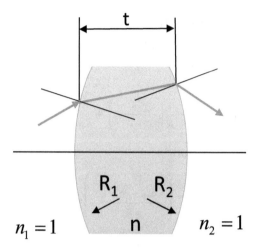

Figure 6.15. Finding the ABCD matrix for a thick lens.

$$\begin{pmatrix} x_2 \\ \theta_2 \end{pmatrix} = M_2 T M_1 \begin{pmatrix} x_1 \\ \theta_1 \end{pmatrix}$$
$$= M \begin{pmatrix} x_1 \\ \theta_1 \end{pmatrix},$$

(6.43)

where M_1 is the matrix for refraction at surface 1, M_2 at surface 2, and T is the translation (propagation) matrix. The order of the matrix multiplication is very important (matrix multiplication is *not* commutative). If equation (6.40) is placed below figure 6.15, it may seem that the matrices appear multiplied in reverse order. However, this is the correct order and one way to remember it is to think of it as representing the *chronological* order of events: first, the input vector is modified by the refraction 1, then by the propagation matrix, etc.

Using the refraction matrix derived in section 6.6.3 and the translation matrix from section 6.6.1, we find

$$M_1 = \begin{pmatrix} 1 & 0 \\ -\dfrac{n-1}{nR_1} & \dfrac{1}{n} \end{pmatrix}$$

(6.44a)

$$T = \begin{pmatrix} 1 & t \\ 0 & 1 \end{pmatrix}$$

(6.44b)

$$M_2 = \begin{pmatrix} 1 & 0 \\ -\dfrac{1-n}{-R_2} & n \end{pmatrix},$$

(6.44c)

where for M_2, we took into account that $R_2 < 0$. The *convergence* of the surface is defined as the refractive index contrast divided by the radius of curvature.

$$C_1 = \frac{n-1}{R_1} \tag{6.45a}$$

$$C_2 = \frac{n-1}{R_2} \tag{6.45b}$$

We see in equation (6.45b) that the negative refractive index contrast at the second surface is cancelled by the negative curvature. Thus, the matrix for the thick lens becomes

$$M = \begin{pmatrix} 1 & 0 \\ -C_2 & n \end{pmatrix} \begin{pmatrix} 1 & t \\ 0 & 1 \end{pmatrix} \begin{pmatrix} 1 & 0 \\ -\dfrac{C_1}{n} & \dfrac{1}{n} \end{pmatrix}$$

$$= \begin{pmatrix} 1 - C_1\dfrac{t}{n} & \dfrac{t}{n} \\ -C_1 - C_2 + C_1 C_2\dfrac{t}{n} & 1 - C_2 t \end{pmatrix}. \tag{6.46}$$

By definition, the lower-left element of the ABCD matrix (i.e., the C-element) in equation (6.46) gives the focal distance of the lens, as

$$\frac{1}{f} = C_1 + C_2 - C_1 C_2\frac{t}{n}. \tag{6.47}$$

Equation (6.47) is known as the lens maker equation for *thick* lenses, as it provides a recipe for creating a lens of the desired focal distance. Note that, if the lens is places in a medium of refractive index n, the focal distance becomes f/n.

6.6.5 Transmission through a thin lens

The matrix for a thin lens is obtained from equation (6.46) by simply taking the limit $t \to 0$ (figure 6.16)

$$M = \begin{pmatrix} 1 & 0 \\ -\dfrac{1}{f} & 1 \end{pmatrix}. \tag{6.48}$$

A horizontal ray passing through a thin lens intersects the optical axis at the focal plane image, F'. Propagating a ray from right to left through the same lens intersects the optical axis at the focal plane object, F (figure 6.16(a)). For the negative lens in figure 6.16(b), a horizontal ray is bent away from the optical axis. The ray continuations (dashed lines in figure 6.16) intersect the optical axis to define *virtual* focal planes. Note that F and F' are in reversed for a negative lens versus a positive one.

Combining lens and translation matrices allows us to describe efficiently complex optical systems, especially in imaging. We will discuss these applications in detail in volume 9 [6].

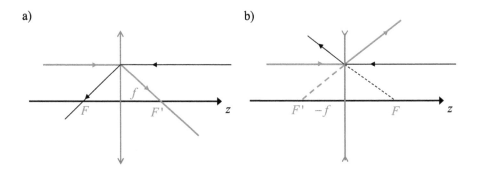

Figure 6.16. (a) A thin, positive lens ($f > 0$). (b) A thin, negative lens ($f < 0$).

6.6.6 Reflection by a spherical mirror

As discussed earlier, mirrors can be used as lenses, with the added benefit of focusing light without chromatic aberrations. The matrix for a spherical mirror is the same as for a thin lens, if we replace the focal distance with $R/2$.

$$M = \begin{pmatrix} 1 & 0 \\ \dfrac{2}{R} & 1 \end{pmatrix}.$$ (6.49)

According to the sign convention, since the center of curvature is to the left of the surface for a concave mirror, $R < 0$. Similarly, $R > 0$ for a convex mirror.

6.6.7 Cascading optical systems

The power of the ABCD matrix formalism becomes obvious when dealing with complex optical systems, consisting of a cascade of optical elements. Let us consider the optical system that transports a ray from point A to point B (figure 6.17). The ray is first translated by matrix T_1, refracted by R_1, and so on. The output vector is obtained simply by multiplying the matrices in the correct, 'chronological' order,

$$\begin{pmatrix} x_2 \\ \theta_2 \end{pmatrix} = T_1 R_1 T_2 R_2 T_3 R_3 T_4 \begin{pmatrix} x_1 \\ \theta_1 \end{pmatrix}.$$ (6.50)

As a simple example, let us consider the thin lens sandwiched by translations on both sides (figure 6.18(a)). The matrix that describes the ray propagation from plane z_1, to z_2 is

$$M = \begin{pmatrix} 1 & z_2 \\ 0 & 1 \end{pmatrix} \begin{pmatrix} 1 & 0 \\ -\dfrac{1}{f} & 1 \end{pmatrix} \begin{pmatrix} 1 & z_1 \\ 0 & 1 \end{pmatrix}$$

$$= \begin{pmatrix} 1 - \dfrac{z_2}{f} & z_1 + z_2 - \dfrac{z_1 z_2}{f} \\ -\dfrac{1}{f} & 1 - \dfrac{z_1}{f} \end{pmatrix}.$$ (6.51)

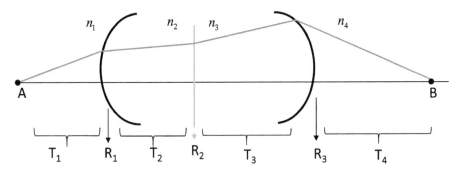

Figure 6.17. Ray propagation through a cascade of optical elements: $T_{1,2,3,4}$ transmission matrices, $R_{1,2,3}$ refraction matrices, $n_{1,2,3,4}$ refractive indices.

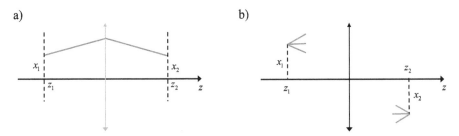

Figure 6.18. (a) Propagation through a thin lens. (b) Imaging condition.

Of particular relevance is element 'B' of the matrix, namely, $z_1 + z_2 - \dfrac{z_1 z_2}{f}$. From the input–output relation, we see that

$$x_2 = Ax_1 + B\theta. \tag{6.52}$$

Thus, in the particular case $B = 0$, the ray elevation, x_2, is independent of the input angle, θ (figure 6.18(b)). This is a very important case as it establishes the condition for *imaging*, i.e., when a point source object is relayed by the imaging system into a single point image. From the matrix in equation (6.51), we see that this imaging condition implies

$$\frac{1}{z_1} + \frac{1}{z_2} = \frac{1}{f}. \tag{6.53}$$

Equation (6.53) describes the relationship between the conjugate planes (z_1, z_2) for a given lens. We will continue this discussion in volume 9 [6], when studying imaging systems.

6.6.8 Eigen vectors

It is informative to discuss the meaning of *eigen vectors* for the matrix of a given optical system. The problem can be expressed in terms of the eigen value λ as

$$\begin{pmatrix} A & B \\ C & D \end{pmatrix} \begin{pmatrix} x \\ \theta \end{pmatrix} = \lambda \begin{pmatrix} x \\ \theta \end{pmatrix}. \tag{6.54}$$

If this equation has solutions, it means that there exist rays that are not modified by the system, except for a scalar multiplication. The condition for non-trivial solutions is

$$\begin{vmatrix} A - \lambda & B \\ C & D - \lambda \end{vmatrix} = 0, \tag{6.55}$$

or,

$$\lambda^2 - (A + D)\lambda + AD - BC = 0, \tag{6.56}$$

which yields the usual solutions, $\lambda_{1,2}$. Note that $AD - BC = 1$, as it equals the determinant of a matrix. A question of particular importance is the following: *is there an input plane for which the output vector is identical to the input and $\lambda = 1$?* If such a solution exists, the input and output planes are called the *principal planes* of the system.

Let us find the principal planes of the system in (6.45): we seek z_1 and z_2 for which $\lambda = 1$. From equation (6.56), $\lambda = 1$, and the matrix elements from equation (6.50) yield

$$1 - \left(1 - \frac{z_1}{f}\right) - \left(1 - \frac{z_2}{f}\right) + 1 = 0, \tag{6.57}$$

which yields $z_1 = z_2$. Since $z_1 z_2 < 0$, the only possible solution is

$$z_1 = z_2 = 0 \tag{6.58}$$

This result indicates that the principal planes for a thin lens overlap with the lens itself. It is left as an exercise to use the same procedure to determine the principal planes of a thick lens (figure 6.19).

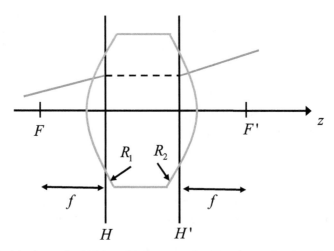

Figure 6.19. Principle planes of a thick lens, H'. A ray crossing H can be translated to H', maintain the same elevation and angle. The focal distances are measured with respect to the principal planes.

Knowing the principal planes offers great simplifications when performing ray tracing through a complex optical system, as a given ray can be transported identically in elevation and angle from one principal plane to another. The principal planes serve as reference with respect to which the focal distances are measured (see volume 9 on imaging [6]). In the next chapter, we will see that the same ABCD matrix formalism introduced here can be used to propagate Gaussian beams as well.

6.7 Problems

1. Write the eikonal equation in the $(\mathbf{k}, \omega)$ representation.
2. Use equation (6.8) to prove that the wavevector is always normal to the wavefront.
3. Use Fermat's principle for refraction at a flat surface to arrive at Snell's law (figure 6.20).
4. Find the focal distance of a concave surface separating media of refractive indices n_1 and n_2. Discuss when the surface acts as a positive and negative lens (figure 6.21).
5. In section 6.4, we calculated the focal distance for a thin, *biconvex* lens. Calculate the focal distance for the following lens types:
    a. Plano-convex
    b. Positive meniscus
    c. Biconcave
    d. Plano-concave
    e. Negative meniscus
6. Calculate the position of the principal planes for the lenses in problem 5 (figure 6.22).
7. Calculate the position of the principal planes of the thick lens shown in figure 6.23.

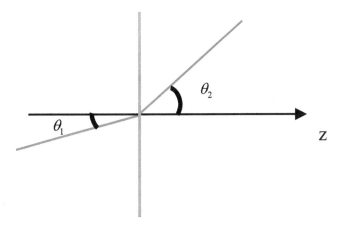

Figure 6.20. Problem 3.

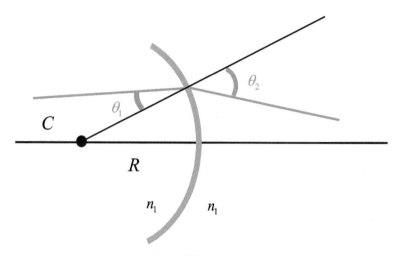

Figure 6.21. Problem 4.

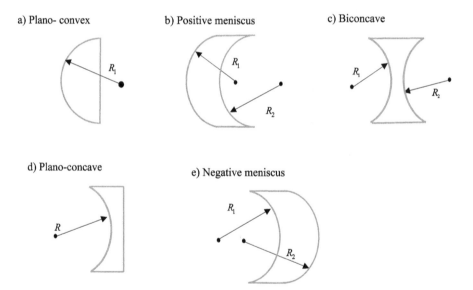

Figure 6.22. Problem 5. Various types of lenses.

8. Calculate the matrix associated with the system shown in figure 6.24.
9. Find the imaging condition for the system in problem 8.
10. Calculate the position of the principal planes for the system in problem 8.
11. Calculate the ABCD matrix that describes the system shown in figure 6.25.
12. Find the imaging condition for the system in figure 6.25.
13. Calculate the positions of the principal planes for the system in figure 6.25.
14. Calculate the ABCD matrix associated with propagating rays between planes z_1 and z_2 (figure 6.26). The $\updownarrow$ symbol indicates thin positive lenses.

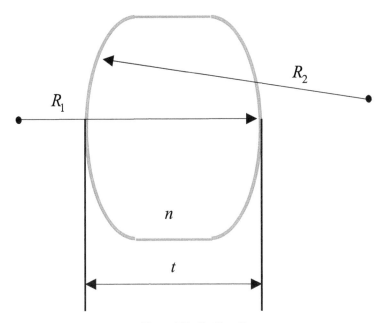

Figure 6.23. Problem 7.

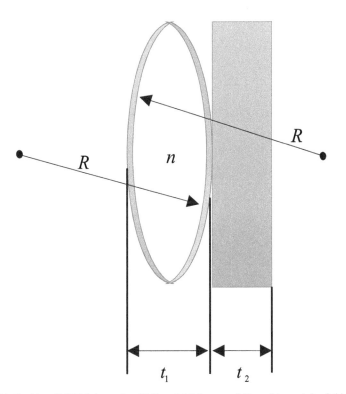

Figure 6.24. Problem 8. Thick lens of radii R and thickness t_1 followed by a slab of thickness t_2.

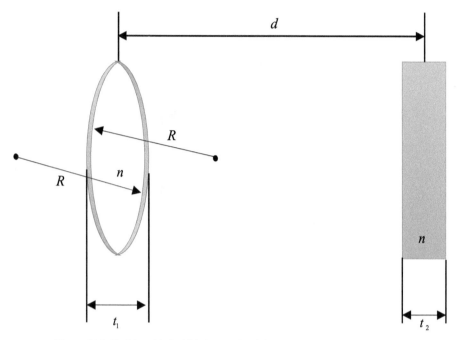

Figure 6.25. Problem 11. A thick lens and a slab are separated by a distance d.

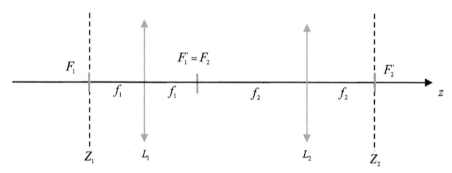

Figure 6.26. Problem 14. $L_{1,2}$ are thin positive lenses, $f_{1,2}$ are the focal distances, $F_{1,2}$ indicate the front focal planes (or focal plane object), and $F'_{1,2}$ are the back focal planes (or focal planes image).

References and further reading

[1] Popescu G *Principles of Biophotonics, Volume 6—Light propagation in Inhomogeneous Media* (Bristol: IOP Publishing) (not yet published)

[2] Popescu G 2018 *Principles of Biophotonics, Volume 1—Linear Systems and the Fourier Transform in Optics* (Bristol: IOP Publishing)

[3] Johnston I 2009 *The Mozi: A Complete Translation* (Hong Kong:: The Chinese University of Hong Kong Press)

[4] Mahoney M S 1994 *The Mathematical Career of Pierre de Fermat, 1601–65* (Princeton, NJ: Princeton University Press) (not yet published)

[5] Newton I 1718 *Opticks* 2nd edn. (London: W. and J. Innys) 4 p.l 382

[6] Popescu G *Principles of Biophotonics, Volume 9—Optical Imaging* (Bristol: IOP Publishing) (not yet published)

[7] Dereniak E L and Dereniak T D 2008 Geometrical and trigonometric optics *Geometrical and Trigonometric Optics* ed E L Dereniak and T D Dereniak (Cambridge: Cambridge University Press)

[8] Fowles G R 1975 *Introduction to Modern Optics* (New York: Holt)

[9] Gerrard A and Burch J M 1975 *Introduction to Matrix Methods in Optics* (London, New York: Wiley)

[10] Synge J L 1937 *Geometrical Optics: an Introduction to Hamilton's Method* (CUP Archive) (Cambridge: Cambridge University Press)

[11] Saleh B E A and Teich M C 2007 *Fundamentals of Photonics* (New York: Wiley)

IOP Publishing

Principles of Biophotonics, Volume 3
Field propagation in linear, homogeneous, dispersionless, isotropic media
Gabriel Popescu

Chapter 7

Gaussian beam propagation

7.1 Definition of a light beam

Many biophotonics instruments use beams of light, e.g., confocal microscopy, optical coherence microscopy, Raman spectroscopy, etc. Let us first define a *beam*, as an optical field that has a dominant wavevector. In other words, a beam propagating along z has a transverse spatial bandwidth much smaller than β_0 (figure 7.1(a) and (b)),

$$\Delta k_x \ll \beta_0 \ . \tag{7.1}$$

Examples of such light include any form of collimated light. The case described in figures 7.1(c) and (d) cannot be considered a beam. Such light can be emitted by extended sources, such as thermal lamps and some LED's. An extreme case of *spatially* broadband light is when the k_x-distribution is uniform. This type of light, of isotropic **k**-distribution, is called *diffusive*, such as that scattered by a projector screen, or the multiply scattered light, deep inside biological tissue.

A beam characterized by $\Delta k_x \ll \beta_0$, is the spatial equivalent of *quasi-mono-chromatic* light, where we have the condition $\Delta\omega \ll \omega_0$, with $\Delta\omega$ the temporal bandwidth, and ω_0 the central frequency. As in the temporal domain, the definition of a beam is not very strict.

In certain cases, the beam profile can be approximated by a Gaussian function which greatly simplifies calculations. A single (spatial) mode laser emits a perfect Gaussian beam. However, under the paraxial approximation other fields may be described by Gaussian beam propagation. In this chapter, we derive the main expression for a Gaussian beam propagation and establish a matrix formalism for efficient calculations, similar to those presented in chapter 6.

7.2 Fresnel propagation of Gaussian beams

In chapter 5, we found that the spherical wave can be approximated by a quadratic phase wavelet, such that the propagation of a field at small angles can be described by the Fresnel approximation,

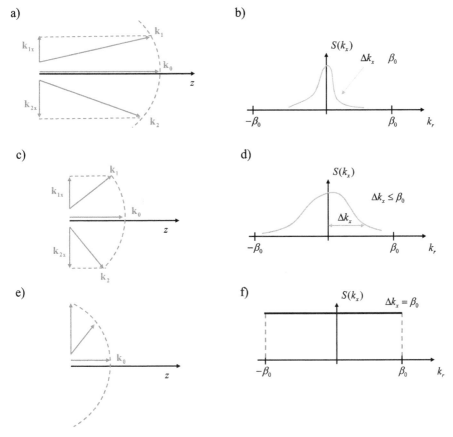

Figure 7.1. (a) k-vector distribution of a *beam*. (b) The transverse power spectrum, $S(k_x)$ is much narrower than the dominant wavenumber, β_0. Optical field that cannot be considered a beam (c) and (d). Diffuse light, characterized by isotropic distribution of k_x (e) and (f).

$$U(\mathbf{r_\perp}, z) = \frac{\beta_0}{z} U(\mathbf{r_\perp}, 0) \circledast_{\mathbf{r_\perp}} e^{i\frac{\beta_0 r_\perp^2}{2z}} \; , \tag{7.2}$$

where we ignore the plane wave pre-factor $e^{i\beta_0 z}$. In the $\mathbf{k_\perp}$-domain, we obtain the Fresnel approximation for the angular spectrum $(\mathbf{k_\perp}, z)$ propagation,

$$U(\mathbf{k_\perp}, z) = U(\mathbf{k_\perp}, 0) e^{-i\frac{z k_\perp^2}{\beta_0}} \tag{7.3}$$

We now consider that the input field, $U(\mathbf{r_\perp}, 0)$, has a Gaussian amplitude (figure 7.2)

$$U(\mathbf{r_\perp}, 0) = A e^{-\frac{r_\perp^2}{w_0^2}} \; , \tag{7.4}$$

where w_0 is called the *beam waist*. The normalization factor, A, is such that $\iint [U(\mathbf{r_\perp}, 0)^2] d^2\mathbf{r_\perp} = 1$. Note that we define the Gaussian amplitude with a width w_0, which, unfortunately is not the standard deviation of the Gaussian distribution. However, this is the traditional definition.

a)

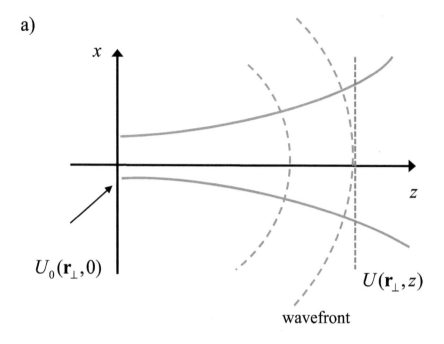

b)

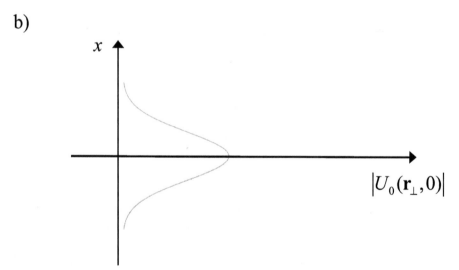

Figure 7.2. (a) Propagation of a Gaussian beam along z. (b) Gaussian amplitude of the input field.

In order to calculate the field distribution at a z-plane, we need to perform the convolution in equation (7.2), which, of course, is best performed in the $\mathbf{k_\perp}$-domain,

$$U(\mathbf{k_\perp}, z) = e^{-\frac{k_\perp^2 w_0^2}{4}} e^{i\frac{zk_\perp^2}{2\beta_0}}$$

$$= e^{-\frac{k_\perp^2}{4}\left(w_0^2 + i\frac{2z}{\beta_0}\right)}., \tag{7.5}$$

where we ignored insignificant pre-factors that do not depend on $\mathbf{k}_\perp$. Note that, as expected, the field maintains a Gaussian shape, except that now the 'width' is complex. Fourier transforming equation (7.5) back to the $\mathbf{r}_\perp$ domain, we obtain

$$U(\mathbf{r}_\perp, z) = \frac{w_0}{c(z)} e^{-\frac{r_\perp^2}{w^2(z)}}, \tag{7.6a}$$

$$c(z) = w_0 \left(1 + i \frac{2z}{\beta_0 w_0^2} \right). \tag{7.6b}$$

Equations (7.6a) and (7.6b) provide the solution for the Gaussian beam propagation along z, with an interesting dependence on the function $c(z)$. Next, we describe in detail the physical interpretation of this solution.

7.3 Gaussian beam characteristics

The prefactor $c(z)$ in equation (7.6) can be expressed in terms of a magnitude, $|c(z)| = w(z)$, and phase, ψ,

$$c(z) = w_0 \left(1 + i \frac{2z}{\beta_0 w_0^2} \right)^{1/2} \tag{7.7a}$$

$$= w(z) e^{i\psi(z)}$$

where, the new parameters are defined as

$$w(z) = w_0 \left[1 + \left(\frac{z}{z_0} \right)^2 \right]^{1/2} \tag{7.7b}$$

$$\psi(z) = \arg\left(\frac{z}{z_0} \right), \tag{7.7c}$$

with

$$z_0 = \frac{\beta_0 w_0^2}{2} \tag{7.7d}$$

$$= \frac{\pi w_0^2}{\lambda}$$

The expression for the Gaussian beam can be re-written as (equation (7.6))

$$U(\mathbf{r}_\perp, z) = \frac{w_0}{w(z)} e^{-\frac{r_\perp^2}{w_0^2}\left[\frac{1}{1+i\frac{z}{z_0}} \right]} e^{-i\psi} e^{i\beta_0 z} \tag{7.8}$$

$$= \frac{w_0}{w(z)} e^{-\frac{r_\perp^2}{w^2(z)}} e^{i\phi(\mathbf{r}_\perp, z)}$$

In equation (7.8), we see that the beam maintains a transverse Gaussian profile, with a z-dependent width, $w(z)$. The phase of the beam, $\phi(\mathbf{r}_\perp, z)$, can be written as

$$\phi(\mathbf{r}_\perp, z) = \beta_0 z + \frac{\beta_0 r_\perp^2}{2R(z)} - \tan^{-1}\left(\frac{z}{z_0}\right),\tag{7.9}$$

where

$$R(z) = z\left(1 + \frac{z_0^2}{z^2}\right).\tag{7.10}$$

Let us now discuss the physical significance of the various parameters contained in the field $U(\mathbf{k}_\perp, z)$. The distance z_0 is called the *Rayleigh range*, i.e., the distance of which the beam width increases by $\sqrt{2}$, (figure 7.3)

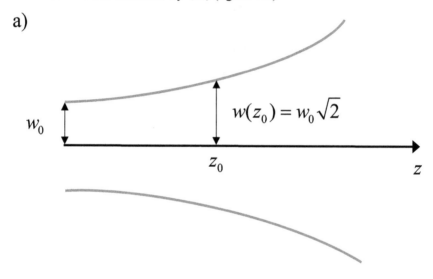

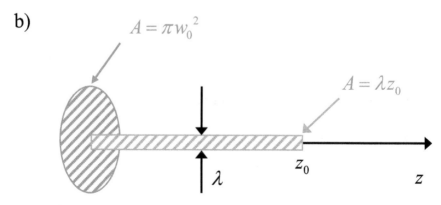

Figure 7.3. (a) The Rayleigh parameter. (b) The beam cross section area equals that of a rectangle formed by z_0 and the wavelength λ.

$$w(z_0) = w_0\sqrt{2} \ . \tag{7.11}$$

One way to remember the formula for z_0 is to realize that, multiplied by the wavelength, it gives the transverse area of the beam, $z_0\lambda = \pi w_0^2$ (figure 7.3). As the beam waist increases, the beam divergence decreases, which makes z_0 larger. For the same beam waist, light of shorter wavelengths diverges more slowly, $z_0 \propto 1/\lambda$. This is consistent with the diffraction results discussed in chapter 5.

In the expression for the phase of the beam, $R(z)$, is the *wavefront radius* of curvature (figure 7.4). Because the Gaussian (parabolic) wavefront is flatter than a spherical wavefront, its center is located to the left of the origin by a *z-dependent* distance $d(z)$,

$$d(z) = R(z) - z$$
$$= \frac{z_0^2}{z}. \tag{7.12}$$

As expected, for very large distances, $d(z) \simeq 0$, and $R(z) \simeq z$.

The beam divergence angle is defined in the *far zone* as (figure 7.5),

$$\theta = \lim_{z \to \infty} \frac{w(z)}{z}.$$
$$\simeq \frac{w_0}{z_0} \tag{7.13}$$

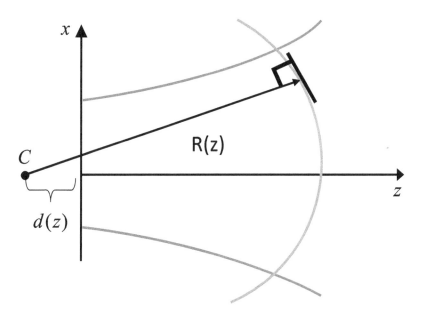

Figure 7.4. Wavefront radius of curvature.

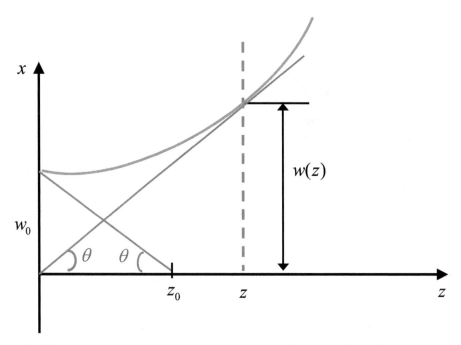

Figure 7.5. Divergence angle of a Gaussian beam is equal to the angle subtended by the beam waist from the Rayleigh range point on axis.

Perhaps the most surprising result is the phase delay due to the last term in equation (7.9), $\psi(z) = \arg(z/z_0)$. It can be seen that on propagating from $z = 0$ to ∞, the wavefront is advanced by $\pi/2$. Overall, from $-\infty$ to ∞, there is a total phase shift of π (figure 7.6). This peculiar phase shift is referred to as *Gouy's phase*, after its discoverer. This phase advancement contributes to the overall phase velocity v_p along the z-axis ($\mathbf{r}_\perp = 0$), which exceeds the speed of light in a vacuum,

$$v_p = \frac{\omega}{k_{eff}}, \tag{7.14}$$

where, k_{eff} is the *effective* wavenumber along the z-axis, defined as

$$k_{eff} = \left[\frac{d}{dz} [k_0 z - \psi(\mathbf{r}_\perp, z)] \right]_{\mathbf{r}_\perp = 0}$$
$$= k_0 - \frac{1/z_0}{1 + \frac{z^2}{z_0^2}} < k_0 \tag{7.15}$$

Thus, the contribution of the Gouy phase is always to advance the phase and bring the phase velocity to superluminal values.

Next, we establish a formalism for calculating the beam properties upon interaction with various optical components.

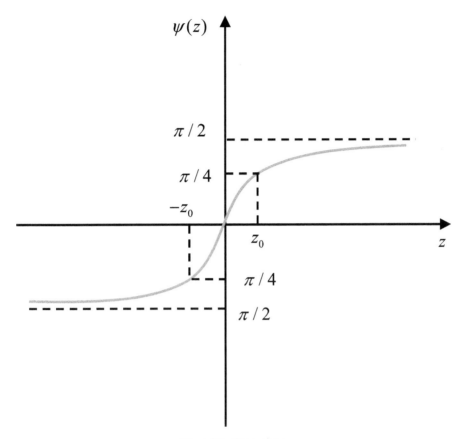

Figure 7.6. Gouy phase.

7.4 Gaussian beam propagation using ABCD matrices

A Gaussian beam is fully determined by two parameters, the width $w(z)$, and the radius $R(z)$. These two quantities are typically combined into a single complex function, referred to as the *complex beam parameter*, q, defined as

$$\frac{1}{q(z)} = \frac{1}{R(z)} + i\frac{\lambda}{\pi w^2(z)} \tag{7.16}$$

Note that, in geometrical optics, rays are fully determined by two parameters as well. In section 6.6 we found that the ABCD matrix formalism provides a practical means for ray propagation. Interestingly, the same matrices can be used to propagate Gaussian beams, as both regimes operate under the paraxial approximation (see section 2.3 in [1]). The input–output relationship for a given optical system characterized by the matrix ABCD is in terms of the q-parameters (figure 7.7),

$$q_2 = \frac{Aq_1 + B}{Cq_1 + D} \tag{7.17a}$$

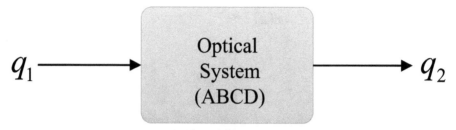

Figure 7.7. Propagation of Gaussian beams as a linear system operation.

or

$$\frac{1}{q_2} = \frac{C + \dfrac{D}{q_1}}{A + \dfrac{B}{q_1}} \tag{7.17b}$$

Let us consider several cases of common interest, which were also discussed in the context of ray optics in section 6.6.

7.4.1 Free space propagation

The free space propagation matrix has a particularly simple form (figure 7.8).

$$\begin{pmatrix} A & B \\ C & D \end{pmatrix} = \begin{pmatrix} 1 & d \\ 0 & 1 \end{pmatrix}. \tag{7.18}$$

As a result, the complex beam parameters output reads

$$q_2 = q_1 + d, \tag{7.19a}$$

or

$$w(z + d) = w_0 \sqrt{1 + \frac{(z + d)^2}{z_0^2}}$$

$$= \frac{w(z) \sqrt{1 + \dfrac{(z + d)^2}{z_0^2}}}{\sqrt{1 + \dfrac{z^2}{z_0^2}}} \tag{7.19b}$$

$$R(z + d) = (z + d) \left[1 + \frac{z_0^2}{(z + d)^2} \right]. \tag{7.19c}$$

From equations (7.19), we see that, at large distances, $z \gg z_0$, neglecting the 1 under the square root,

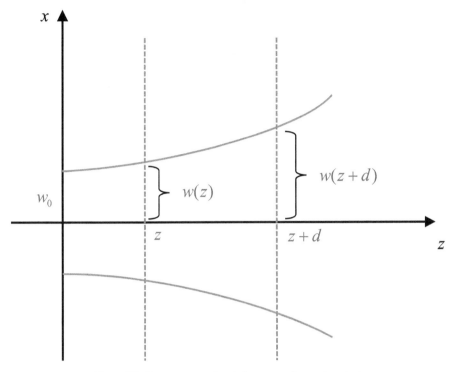

Figure 7.8. Beam propagation in free space from z to $z + d$.

$$w(z + d) \simeq w(z)\frac{z + d}{z} \ . \tag{7.20}$$

Finally, if the input happens to coincide with the beam waist, $z = 0$, then

$$q_1(0) = -i\frac{\pi w_0^2}{\lambda} \tag{7.21a}$$

and

$$q_2(d) = d - i\frac{\pi w_0^2}{\lambda}, \tag{7.21b}$$

or

$$\frac{1}{q_2(d)} = -\frac{d}{d^2 + \left(\frac{\pi w_0^2}{\lambda}\right)^2} + i\frac{\pi w_0^2}{\lambda\left[d^2 + \left(\frac{\pi w_0^2}{\lambda}\right)^2\right]} \tag{7.21c}$$

$$= \frac{1}{R_2(d)} + i\frac{\pi w^2(d)}{\lambda}.$$

Equations (7.21a)–(7.21c) yield

$$R_2(d) = d\left[1 + \left(\frac{\pi w_0^2}{\lambda}\right)^2 \frac{1}{d^2}\right]$$

(7.22a)

$$= d(1 + \frac{z_0^2}{d^2})$$

$$w(d) = w_0\sqrt{1 + \left(\frac{d}{z_0}\right)^2},$$

(7.22b)

Which are consistent with equations (7.7b) and (7.10).

7.4.2 Refraction through a planar interface

The matrix for a planar interface is

$$\begin{pmatrix} A & B \\ C & D \end{pmatrix} = \begin{pmatrix} 1 & 0 \\ 0 & \frac{n_1}{n_2} \end{pmatrix}.$$

(7.23)

The q-parameter right after the interface is

$$q_2 = \frac{n_2}{n_1}q_1,$$

(7.24)

Importantly, the wavelength is adjusted in the medium of refractive index $n_{1,2}$ to $\lambda/n_{1,2}$. Thus, equation (7.23) implies

$$R_2(z) = \frac{n_2}{n_1}R_1(z)$$

(7.25a)

$$w_2(z) = w_1(z).$$

(7.25b)

While the beam width remains the same as expected, the radius of curvature is changed (figure 7.9).

7.4.3 Refraction through a spherical interface

The matrix for a spherical surface is

$$\begin{pmatrix} A & B \\ C & D \end{pmatrix} = \begin{pmatrix} 1 & 0 \\ -\frac{1}{f} & \frac{n_1}{n_2} \end{pmatrix},$$

(7.26a)

where

$$\frac{1}{f} = \frac{n_2 - n_1}{n_1 R}$$

(7.26b)

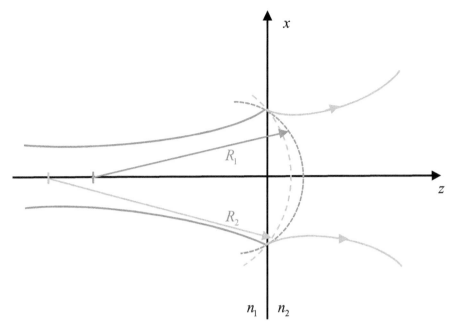

Figure 7.9. The radius of curvature changes at a planar interface.

Therefore, the q-parameter after refraction is

$$q_2(z) = \frac{q_1}{-\dfrac{q_1}{f} + \dfrac{n_1}{n_2}} \qquad (7.27a)$$

$$\begin{aligned}
\frac{1}{q_2(z)} &= -\frac{1}{f} + \frac{n_1}{n_2} \cdot \frac{1}{q_1} \\
&= -\frac{1}{f} + \frac{n_1}{n_2 R_1(z)} + i\frac{\lambda}{\pi w_1^2(z)} \qquad (7.27b) \\
&= \frac{1}{R_2(z)} + i\frac{\lambda}{\pi w_1^2(z)}
\end{aligned}$$

Again, the surface does not change the beam size, only the radius of curvature,

$$w_2(z) = w_1(z_1) \qquad (7.28a)$$

$$R_2(z) = \frac{n_2 f}{n_1 f - n_2 R_1(z)} R_1(z) \qquad (7.28b)$$

Interestingly, the denominator in equation (7.28b) can be positive or negative, depending on whether the lens maintains or changes the sign curvature of the beam. When $n_1 f = n_1 R_1(z)$, the lens perfectly collimates the beam (figure 7.10).

a) $n_1 f > n_2 R_1(z)$

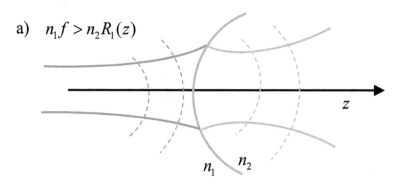

b) $n_1 f = n_2 R_1(z)$

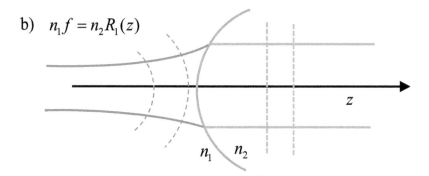

c) $n_1 f < n_2 R_1(z)$

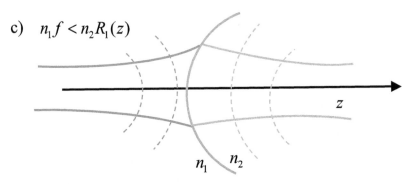

Figure 7.10. Change of beam curvature at a spherical interface. (a) Curvature sign is maintained. (b) Beam is collimated ($R_2 \to \infty$). (c) Curvature is sign changed (divergent to convergent beam, in this case).

7.4.4 Transmission through a thin lens

The matrix for a thin lens is

$$\begin{pmatrix} A & B \\ C & D \end{pmatrix} = \begin{pmatrix} 1 & 0 \\ -\dfrac{1}{f} & 1 \end{pmatrix}, \tag{7.29}$$

where f is the focal distance.

Thus,

$$q_2 = \frac{1}{1 - \dfrac{q_1}{f}}, \tag{7.30}$$

and we can use the result in equation (7.28), for $n_1 = n_2 = 1$, namely

$$w_2(z) = w(z_1) \tag{7.31a}$$

$$R_2(z) = \frac{f}{f - R_1(z)} R_1(z) \tag{7.31b}$$

As before, depending on the sign of $[f - R_1(z)]$, the lens can change the curvature of the beam or make it flat ($R_2 \to \infty$), i.e., collimate the beam (figure 7.11).

It is left as an exercise to study the beam propagation through a thick lens.

7.4.5 Reflection by a spherical mirror

The results from the thin lens can be readily applied here, by replacing f with $R/2$.

7.4.6 Cascading optical systems

As discussed in the ray optics chapter 6, cascading multiple optical systems can be described very effectively using the ABCD matrix formalism (figure 7.12). In order to calculate the beam parameters, we multiply the individual matrices in the correct 'chronological' order

$$\begin{pmatrix} A & B \\ C & D \end{pmatrix} = M_n M_{n-1} \ldots M_1. \tag{7.32}$$

7.5 Problems

1. At a plane $z = 0$, a beam has the following profile $U(x, y, 0) = Ae^{-(\frac{x^2}{w_x^2} + \frac{y^2}{w_y^2})}$. Calculate the phase and amplitude of the field at a plane z. Define the beam divergence, Rayleigh range, and wavefront curvature.
2. A Gaussian beam of divergence $\theta_1 = 1$ mrad is focused down by a lens to half its waist at a distance $d = 1$ m (figure 7.13). What is the focal distance of the lens?
3. A Gaussian beam was measured at two different planes, z_1 and z_2, and the two beam widths were w_1 and w_2. Calculate the beam waist and is location with respect to z_1 (figure 7.14).
4. A Gaussian beam is focused in air by a lens of focal distance f into a spot of waist w_0 and divergence θ_0. What are the waist and divergence of the beam if focused at a depth h in water (figure 7.15)? Discuss the dependence on h, for $0 \leqslant h \leqslant f$.

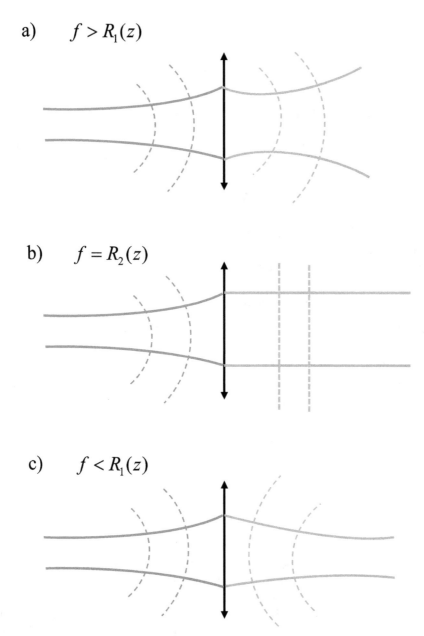

a) $f > R_1(z)$

b) $f = R_2(z)$

c) $f < R_1(z)$

Figure 7.11. Beam transmission through a thin lens. (a) Curvature sign is maintained. (b) Beam is collimated. (c) Curvature sign is changed (divergent to convergent, in this case).

5. A beam of waist w_0 and divergence θ_0 refracts at an interface between air and water. Calculate the new beam waist and divergence at the plane z behind the interface, if the location of the waist of the input beam is z_0 (Rayleigh range) in front of the interface (figure 7.16).

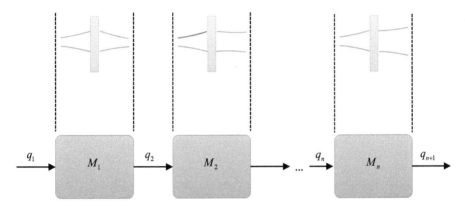

Figure 7.12. Cascading optical systems.

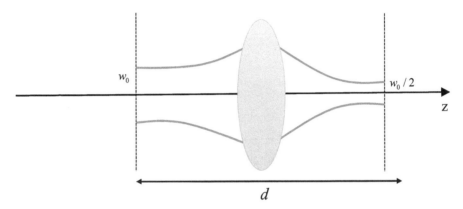

Figure 7.13. Problem 2.

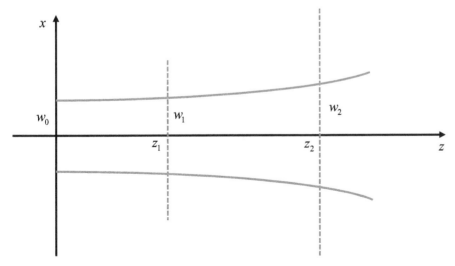

Figure 7.14. Problem 3.

a)   b)

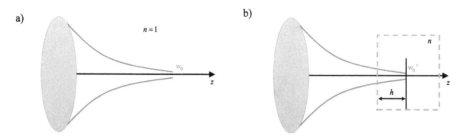

Figure 7.15. . Problem 4: (a) focusing in air; (b) focusing in water.

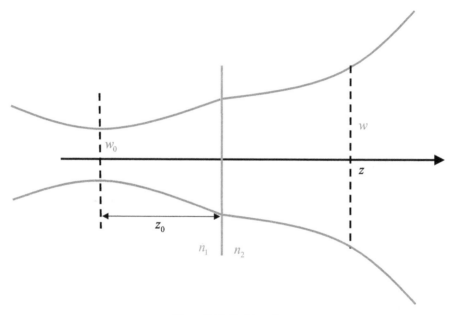

Figure 7.16. Problem 5.

6. A Gaussian beam of waist w_0 and divergence θ_0 passes through the *telecentric* system, as shown in figure 7.5. If the location of the input beam waist is at the front focal plane of lens L_1, calculate the beam width and radius at the back focal plane of lens L_2 (figure 7.17).

7. A beam of width w and wavefront radius R is incident on a concave mirror of radius R_1 as shown in figure 7.18. Calculate the width w' and wavefront radius R' of the reflected beam at a distance z from the mirror.

8. Redo problem 7 with a flat mirror.

9. Redo problem 7 when the wavefront beam matches the mirror curvature.

10. Redo problem 7 when the concave mirror is replaced by a convex one.

11. Redo problem 7, with the mirror replaced by a positive lens of focal distance $R_1/2$.

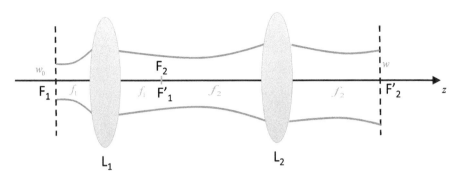

Figure 7.17. Problem 6: Gaussian beam propagation through a telecentric system, characterized by a system of lenses with overlapping focal planes, as indicated.

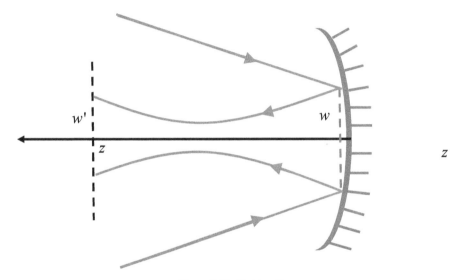

Figure 7.18. Problem 7.

References and further reading

[1] Yariv A and Yeh P 2003 *Optical Waves in Crystals: Propagation and Control of Laser Radiation* (Wiley Classics Library) (Hoboken, NJ: Wiley) xi 589
[2] Saleh B E A and Teich M C 2007 *Fundamentals of Photonics* (New York: Wiley)
[3] Chang W S C 2005 *Principles of Lasers and Optics* (Cambridge, New York: Cambridge University Press)
[4] Siegman A E 1986 *Lasers* (Mill Valley, CA: University Science Books)
[5] Laufer G 1996 *Introduction to Optics and Lasers in Engineering* (Cambridge, New York: Cambridge University Press)

IOP Publishing

Principles of Biophotonics, Volume 3
Field propagation in linear, homogeneous, dispersionless, isotropic media
Gabriel Popescu

Chapter 8

Propagation of field correlations

8.1 Heisenberg uncertainty relation and the coherence of light

8.1.1 Uncertainty relations in space and time

The inherent uncertainty in the optical fields encountered in nature is captured most succinctly by Heisenberg's uncertainty relation. In volume 1 chapter 8 [1], we discussed this uncertainty for arbitrary signals and derived it from the Schwartz inequality [1]. The Heisenberg uncertainty relation states that the spread of a signal in time (space) and temporal (spatial) frequency cannot be arbitrarily small, namely,

$$\Delta t \ \Delta \omega \geqslant 1/2 \qquad (8.1a)$$

$$\Delta x \ \Delta k_x \geqslant 1/2 \qquad (8.1b)$$

$$\Delta y \ \Delta k_y \geqslant 1/2 \qquad (8.1c)$$

$$\Delta z \ \Delta k_z \geqslant 1/2 \qquad (8.1d)$$

The spreads denoted by Δ are the *standard deviations* of the distributions in their respective domains. For an optical field $U(\mathbf{r}, t)$ and its spatiotemporal Fourier transform $U(\mathbf{k}, \omega)$, defined as,

$$U(\mathbf{r}, t) \overset{\mathbf{r}, \ t}{\longleftrightarrow} U(\mathbf{k}, \omega), \qquad (8.2)$$

doi:10.1088/978-0-7503-1646-0ch8

the temporal quantities in equations (8.1a)–(8.1d) are defined as (figure 8.1(a) and (b))

$$\Delta t^2 = \frac{\int_{-\infty}^{\infty} [t - <t>]^2 |\, U(t)|^2 \, dt}{\int_{-\infty}^{\infty} |\, U(t)|^2 \, dt} \tag{8.3a}$$

$$\Delta \omega^2 = \frac{\int_{-\infty}^{\infty} [\omega - <\omega>]^2 |\, U(\omega)|^2 \, d\omega}{\int_{-\infty}^{\infty} |\, U(\omega)|^2 \, d\omega}, \tag{8.3b}$$

In equation (8.3), the average quantities, $<t>$ and $<\omega>$, are defined also with respect to the $|\, U(t)|^2$ and $|\, U(\omega)|^2$ distributions, i.e., the instantaneous irradiance and power spectrum, respectively.

The spatial domain inequalities are defined in full analogy to equation (8.3)

$$\Delta x^2 = \frac{\int \int \int_{-\infty}^{\infty} [x - <x>]^2 |\, U(\mathbf{r})|^2 \, d^3\mathbf{r}}{\int \int \int_{-\infty}^{\infty} |\, U(\mathbf{r})|^2 \, d^3\mathbf{r}} \tag{8.4a}$$

$$\Delta k_x^2 = \frac{\int \int \int_{-\infty}^{\infty} [k_x - <k_x>]^2 |\, U(\mathbf{k})|^2 \, d^3\mathbf{k}}{\int \int \int_{-\infty}^{\infty} |\, U(\mathbf{k})|^2 \, d^3\mathbf{k}}, \tag{8.4b}$$

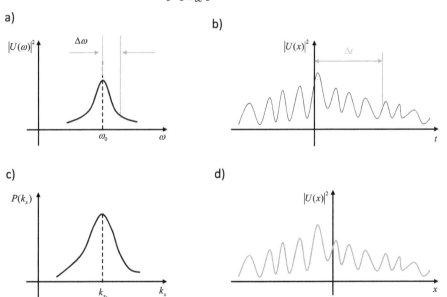

Figure 8.1. (a) Temporal power spectrum of mean frequency ω_0. (b) Temporal distribution of the signal characterized by the power spectrum in (a). (c) Spatial power spectrum along x (the distribution is over the x component of the wavevector, k_x). (d) The superposition of the wave vectors in (c) give the spatial distribution.

while Δy^2, Δk_y^2, Δz^2, Δk_z^2, are defined analogously. The distribution $| U(\mathbf{r})|^2$ is the *local* irradiance, instantaneous, or averaged, depending on whether there is averaging in the time domain. Of course, $| U(k)|^2$ is the spatial power spectrum.

The inequalities in equation (8.1) establish a fundamental fact, perhaps under-appreciated in classical optics: *all* optical fields encountered in nature are subject to uncertainty and thus, are best described by statistics (figure 8.1).

As discussed in volume 1, chapter 8, there are four immediate implications of the uncertainty relations [1]:

(i) Generating small Δt, i.e., short light pulses, requires large temporal bandwidth, $\Delta \omega$.
(ii) Hight spectral resolution, i.e., small $\Delta \omega$, requires large signal spread in time, Δt (for example, in a laser cavity, the longer the light travels inside the resonator, the narrower the emission bandwidth).
(iii) Confining light to small regions of space, Δx, requires large spatial bandwidth, Δk_x (e.g., high resolution imaging requires high numerical apertures).
(iv) Collimating a light beam, i.e., reducing Δk_x, requires broad beams, large Δx.

As proven in section 8.2 of volume 1 [1], the equality in equation (8.1), is achieved for Gaussian distributions of irradiances,

$$| U(t)|^2 = U_0 e^{-\frac{t^2}{2(\Delta t)^2}}$$
$$| U(x)|^2 = U_0 e^{-\frac{x^2}{2(\Delta x)^2}}$$

(8.5)

8.1.2 Uncertainty relation and the Wiener–Khintchine theorem

Due to the Wiener–Khintchine theorem, the uncertainty relations have direct consequences to the temporal and spatial field correlations, as they are the Fourier transform of the respective power spectra, namely,

$$\Gamma(\tau) \leftrightarrow S(\omega) = | U(\omega)|^2$$ (8.6a)

$$W(\rho) \leftrightarrow P(\mathbf{k}) = | U(\mathbf{k})|^2$$ (8.6b)

In equation (8.6), Γ and W are the temporal and spatial autocorrelation functions, respectively, and the fields were assumed *stationary* and *statistically homogeneous*, at least in the wide sense, as defined in volume 1, chapter 8, and revisited in section 8.2 below [1].

According to the scaling theorem of Fourier transforms, the spread of the power spectra and autocorrelations are inversely proportional,

$$\Delta \tau \propto \frac{1}{\Delta \omega}$$ (8.7a)

$$\Delta \rho \propto \frac{1}{\Delta \mathbf{k}} = \frac{\Delta \mathbf{k}}{| \Delta \mathbf{k} |^2}$$ (8.7b)

where $\Delta\tau$ is a measure of temporal coherence and $\Delta\rho$ of spatial correlations ($\Delta\rho \in \mathfrak{R}^3$). The quantities $\Delta\omega$ and Δk represent the spreads of the power spectrum and, if defined as standard deviations, are identical to the ones in equation (8.1).

However, the quantities $\Delta\tau$ and $\Delta\rho$ should not be confused with the temporal and spatial spreads of the signals, Δt, Δx, etc shown in equation (8.1). For example, for a given power spectrum $S(\omega)$, the temporal autocorrelation Γ is uniquely given by its Fourier transform and, thus, its spread, $\Delta\tau$, is also uniquely established. On the other hand, for a fixed power spectrum, the spread of temporal signal itself (standard deviation of the irradiance), Δt, can have any value, provided that (equation (8.1a))

$$\Delta t \geqslant \frac{1}{2\Delta\omega}. \tag{8.8}$$

The reason for such a qualitative difference between $\Delta\tau$ and Δt is that, for a given power spectrum, the field can have an infinite number of spectral phase signals, $\phi(\omega)$, which can yield different Δt (spread) values but the same $\Delta\tau$ (correlation time) values. To make this point perfectly clear, we recall that on page 8-6 of volume 1 [1], we found an expression for the temporal variance in terms of the spectral amplitude and phase,

$$\Delta t^2 = \frac{1}{2\pi}\int_{-\infty}^{\infty}\left[\left(\frac{dA(\omega)}{d\omega}\right)^2 + A^2(\omega)\left(\frac{d\phi(\omega)}{d\omega}\right)^2\right]d\omega - <t>^2. \tag{8.9a}$$

We can readily see from equation (8.9a) that, for a frequency domain field, $U(\omega) = A(\omega)e^{i\phi(\omega)}$ of fixed spectrum, $S(\omega) = A^2(\omega)$, there are infinitely many spreads, Δt, depending on the spectral phase $\phi(\omega)$. How dispersive materials affect $\phi(\omega)$ and, thus, the pulse spread, Δt, and, more importantly, how to perform dispersion compensation, is subject to current research in many fields, including biophotonics. We will discuss in more detail the propagation of light in dispersive media in volume 5 [2].

Of course, the discussion translates analogously to the spatial domain,

$$\Delta x^2 = \frac{1}{2\pi}\int_{-\infty}^{\infty}\left[\left(\frac{dA(k_x)}{dk_x}\right)^2 + A^2(k_x)\left(\frac{d\phi(k_x)}{dk_x}\right)^2\right]dk_x - <x>^2. \tag{8.9b}$$

In 'scanning' imaging systems, i.e., when the image is formed by scanning a focused beam, Δx can be a good measure of spatial resolution. The functional dependence of $\phi(k_x)$ describes how this spatial resolution is degraded. Various terms in a $\phi(k_x)$ power series define particular types of geometric abberations (e.g., spherical aberration, coma, etc). A common expansion of $\phi(k_x)$ is in terms of Zernike polynomials. Imaging systems and aberrations will be discussed in more detail in volume 9 [3].

We note that, because of the Fourier relationship between the power spectrum and the autocorrelation function, we can derive analogous uncertainty relations involving standard deviations of the distributions $|S(\omega)|^2$, $|\Gamma(\tau)|^2$, $|P(k_x)|^2$, $|W(\rho_x)|^2$,

$$\Delta\tau'^2 = \int_{-\infty}^{\infty} (\tau - \langle\tau\rangle)^2 \mid \Gamma(\tau)\mid^2 d\tau \tag{8.10a}$$

$$\Delta\omega'^2 = \int_{-\infty}^{\infty} (\omega - \langle\omega\rangle)^2 \mid S(\omega)\mid^2 d\omega \tag{8.10b}$$

$$\Delta\rho_x'^2 = \int\int\int_{-\infty}^{\infty} \left(\rho_x - \langle\rho_x\rangle\right)^2 \mid W(\boldsymbol{\rho})\mid^2 d^3\boldsymbol{\rho} \tag{8.10c}$$

$$\Delta k_x'^2 = \int\int_{-\infty}^{\infty}\int (k_x - \langle k_x\rangle)^2 \mid P(\mathbf{k})\mid^2 d^3\mathbf{k} \tag{8.10d}$$

where we assume that all distributions are normalized to become probability densities. With these new standard deviations, we can express the uncertainty relations.

$$\Delta\tau'\Delta\omega' \geqslant 1/2 \tag{8.11a}$$

$$\Delta\rho_x'\Delta k_x' \geqslant 1/2 \tag{8.11b}$$

Although these spreads are not commonly used in practice, it is useful to remember that such uncertainties exist for the spreads in terms of spectra and correlations.

8.1.3 Uncertainty relations and diffraction of light

From the uncertainty relations, we see that the only fully determined field is the *monochromatic plane wave*, $e^{-i(\omega_0 t - \mathbf{k}_0 \cdot \mathbf{r})}$, for which $S(\omega) = S_0\delta(\omega - \omega_0)$ and $P(\mathbf{k}) = P_0\delta(\mathbf{k} - \mathbf{k}_0)$. Of course, such a field is infinitely broad in space and time and can never be observed in nature. Thus, all fields encountered in practice have finite spreads in time and space and, therefore, finite spreads in their respective frequency domain.

Interestingly, light diffraction, which was described in chapter 5, can be seen as the manifestation of Heisenberg's uncertainty. Because the spatial spread of the field is constrained by, say, an aperture, the $\mathbf{k}$-vector spread increases accordingly. Let us consider the diffraction of a plane wave at a rectangular aperture (figure 8.2). If we consider the Fraunhofer approximation for *far-zone* diffraction, the diffracted field at plane z is the Fourier transform of the input field,

$$U(x, y, z) = A\int\int_{-\infty}^{\infty} U(x, y, 0)e^{i(k_x x + k_y y)}dxdy,$$
$$= AU(k_x, k_y, 0; \omega)\Big|_{\substack{k_x = \frac{\beta_0 x}{z} \\ k_y = \frac{\beta_0 y}{z}}} \tag{8.12}$$

where the constant A includes unimportant terms for this discussion. Thus, the uncertainty relation applied to the input field at $z = 0$, yields a relationship between the spatial distributions constrained by the aperture and the corresponding spread in $\mathbf{k}$-vectors,

$$\Delta x \; \Delta k_x \geqslant 1/2. \tag{8.13}$$

For a rectangular aperture of width a, the standard deviation Δx can be calculated simply using $\mid U(x)\mid^2 = \dfrac{1}{a}\Pi\left(\dfrac{x}{a}\right)$,

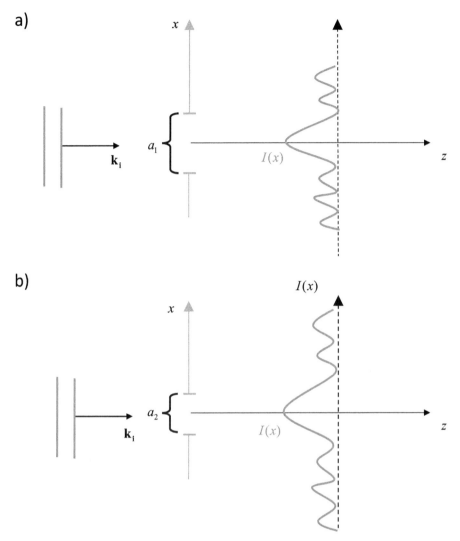

Figure 8.2. Diffraction at an aperture as a manifestation of the uncertainty relation. (a) Aperture of width a_1. (b) Aperture of width $a_2 < a_1$ generates broader intensity distribution, $I(x)$.

$$\Delta x^2 = \frac{1}{a} \int_{-a/2}^{a/2} x^2 dx$$
$$= \frac{x^2}{3a} \Bigg|_{-a/2}^{a/2}$$
$$= \frac{1}{3a} \left(\frac{a^3}{8} + \frac{a^3}{8} \right) \tag{8.14}$$
$$= \frac{a^2}{12}.$$

Thus, the uncertainty relation dictates that at plane z

$$\Delta k_x = \frac{\beta_0}{z} \Delta x'$$
$$\geqslant \frac{\sqrt{3}}{a},$$

(8.15)

where $\Delta x'$ is the standard deviation of the irradiance distribution at plane z. In other words, the spread in x at the output plane satisfies

$$\Delta x' \geqslant \frac{z\sqrt{3}}{\beta_0 a}.$$

(8.16)

Note that the spread at the input plane depends only on the amplitude across the aperture. A variety of spreads $\Delta x'$ at the output plane can be obtained depending on the phase distribution across the aperture. For example, a diffusive phase screen that generates an *isotropic* distribution of **k**-vectors in the semispace $z > 0$ yields $\Delta x' \rightarrow \infty$.

Since we do not deal with a Gaussian distribution, we anticipate that $\Delta k_x > \sqrt{3}/a$. Because diffraction modifies the angular spectrum of light, we expect that the spatial correlations of the field will also be changed. We will discuss the propagation of correlations in the subsequent sections. Next, we refresh the definitions of spatiotemporal correlation functions and their form in certain simplifying cases.

8.2 Spatiotemporal field correlations

8.2.1 Spatiotemporal statistics

From section 8.1, we concluded that a *deterministic*, fully *predictable* field in both time and space is the *monochromatic plane wave*, which, of course, is only a mathematical construct, impossible to obtain in practice due to the uncertainty relation. A thermal source, such as an incandescent filament or the surface of the Sun, emits light in a manner that cannot be predicted with certainty. In other words, unlike in the case of a monochromatic plane wave, we cannot find a function $f(r,t)$ that prescribes the field at each point in space and at each moment in time. Instead, we describe the source as emitting a random signal, $s(r,t)$, and describe its behavior via probability distributions. We can gain knowledge about the random process only by repetitive measurements and averaging the results. This type of averaging over many realizations of a certain random variable is called *ensemble averaging*. The importance of the ensemble averaging has been emphasized in chapter 15, volume 2 [4]. The discipline that studies field correlations is called coherence theory or statistical optics.

Besides its importance to basic science, coherence theory is crucial in predicting outcomes of many light experiments. Whenever we measure a superposition of fields (e.g., in interferometry and imaging) the result of the statistical average performed by the detection process is strongly dependent on the coherence properties of the light. Importantly, half of the 2005 Nobel Prize in Physics was awarded to Roy

Glauber 'for his contribution to the quantum theory of optical coherence.' For a selection of Glauber's seminal papers, see reference [5].

In imaging, because an image is just a (complicated) interferogram [6], the second-order statistics, both in space and time, is of particular interest. Field *cross-correlation functions* are measured whenever we study the relative properties of two fields, one of which is modified by a sample of interest, e.g., in interferometry, holography, and microscopy [7]. Field *auto-correlation functions* are used to describe the properties of the field itself [8], e.g., when studying the statistics of the field upon propagation [9]. Spatial field correlations have been studied in the context of 3D microscopy [8] and quantitative phase imaging [10].

As described in chapter 15, volume 2 [4], a starting point in understanding the physical meaning of a *statistical optical field* is to ask the question: what is the *effective (average) temporal* sinusoid, i.e., $\langle e^{-i\omega t} \rangle_\omega$, for a broadband field? Similarly, what is the *average spatial* sinusoid, i.e., $\langle e^{-i\omega t} \rangle_k$? As before, we use the sign convention whereby a monochromatic plane wave is described by $e^{-i(\omega t - k \cdot r)}$ (see volume 1, section 3.2 [1]). These two averages can be performed using the *probability densities* associated with the temporal and spatial frequencies, $S(\omega)$ and $P(\mathbf{k})$, respectively, which are normalized to satisfy $\int S(\omega) d\omega = 1$ and $\int P(\mathbf{k}) d^3\mathbf{k} = 1$. Thus, $S(\omega) d\omega$ is the probability of having frequencies in the range $(\omega, \omega + d\omega)$ in our field, or the fraction of the total power contained in the vicinity of frequency ω. Similarly, $P(\mathbf{k}) d^3\mathbf{k}$ is the probability of having spatial frequencies in the infinitesimal volume around $\mathbf{k}$, or the fraction of the total power contained around spatial frequency $\mathbf{k}$. Up to a normalization factor, *S and P are the temporal and spatial power spectra* associated with the fields. Thus, the two 'effective sinusoids' can be expressed as *ensemble averages*, using $S(\omega)$ and $P(\mathbf{k})$ as weighting functions,

$$\langle e^{-i\omega t} \rangle_\omega = \int S(\omega) e^{-i\omega t} d\omega$$
$$= \Gamma(t) \tag{8.17a}$$

$$\langle e^{i\mathbf{k} \cdot \mathbf{r}} \rangle_{\mathbf{k}} = \int P(\mathbf{k}) e^{i\mathbf{k} \cdot \mathbf{r}} d^3\mathbf{k}$$
$$= W(\mathbf{r}) \tag{8.17b}$$

Equations (8.17a) and (8.17b) establish that the average *temporal sinusoid* for a broadband field equals its temporal autocorrelation, denoted by Γ. Similarly, the average *spatial sinusoid* for an inhomogeneous field equals its spatial autocorrelation, denoted by W.

Coherence theory can make predictions of experimental relevance. The general problem can be formulated as follows: given the optical field distribution $U(\mathbf{r}, t)$ that varies randomly in space and time, over what *spatiotemporal domain* does the field preserve significant correlations? Experimentally, this question translates into: combining the field $U(\mathbf{r}, t)$ with a replica of itself shifted in both time and space, $U(\mathbf{r} + \boldsymbol{\rho}, t + \tau)$, *on average*, how large can $\boldsymbol{\rho}$ and τ be and still observe 'significant' interference, or obtain interference with 'significant' contrast?

The statistical behavior of an optical field U, assumed to be scalar, can be mathematically captured generally via its *spatiotemporal correlation function*

$$\Lambda(\mathbf{r}_1, \mathbf{r}_2; t_1, t_2) = \langle U(\mathbf{r}_1, t_1)U \times (\mathbf{r}_2, t_2)\rangle, \tag{8.18}$$

where the angle brackets denote *ensemble average*. In essence, this auto-correlation function quantifies how similar the field is with respect to a shifted version of itself, in time or space. To obtain the spatiotemporal correlation function defined in equation (8.18), one needs to average over the product of the samples in both time and space (a total of four summations). Thus, performing ensemble averages is often difficult. However, a simplifying assumption, allows us to compute averages from a single, sufficiently long, sample, if we deal with fields that are both *stationary* (in time) and *statistically homogeneous* (in space), as discussed for generic stochastic signals in volume 1, section 9.6 [1].

Wide-sense stationarity is less restrictive and defines a random process with only its first and second moments independent of the choice of origin. For the discussion here, the fields are assumed to be stationary at least in the wide sense. Under these circumstances, the dimensionality of the spatiotemporal correlation function Λ decreases by half, and we can write

$$\Lambda(\boldsymbol{\rho}, \tau) = \langle U(\mathbf{r}, t)U \times (\mathbf{r} + \boldsymbol{\rho}, t + \tau)\rangle_{\mathbf{r}, t}, \tag{8.19}$$

where $\boldsymbol{\rho} = \mathbf{r}_2 - \mathbf{r}_1$ and $\tau = t_2 - t_1$.

Note that $\Lambda(\mathbf{0,0}) = \langle U(\mathbf{r}, t)U \times (\mathbf{r}, t)\rangle_{\mathbf{r}, t}$ represents the *averaged irradiance* of the field, which is, of course, a real quantity. However, in general $\Lambda(\boldsymbol{\rho}, \tau)$ is complex. Let us define a normalized version of Λ, referred to as the *spatiotemporal complex degree of coherence*

$$\alpha(\boldsymbol{\rho}, \tau) = \frac{\Lambda(\boldsymbol{\rho}, \tau)}{\Lambda(\mathbf{0,0})} \tag{8.20}$$

It can be shown that for stationary fields $|\Lambda|$ attains its maximum at $\mathbf{r} = 0$, $t = 0$, thus

$$0 < |\alpha(\rho, \tau)| < 1. \tag{8.21}$$

Note that, in homogeneous media, such as free space, the *dispersion relation* connects the three coordinates of the wavevector, $k^2 = k_x^2 + k_y^2 + k_z^2 = \omega^2/c^2$. As a consequence, in most cases, the *spatial* behavior of the optical field is well described in 2D, i.e., in a plane. All measurements are performed either by 1D or 2D detectors; thus, we use the coherence *time* and *area* to characterize field statistics. We can define an area $A_C \propto \rho_C^2$ and length $l_c = c\tau_C$, over which $|\alpha(\rho_C, \tau_C)|$ maintains a significant value, say $|\alpha| > 1/2$, which defines a *coherence volume*

$$V_c = A_c\, l_c. \tag{8.22}$$

This *coherence volume* determines the maximum domain size over which the fields can be considered correlated. These parameters are of practical importance because they indicate over what spatiotemporal domain a field distribution maintains significant correlation with respect to a shifted replica of itself.

Generally, the random signal, $s(\mathbf{r}, t)$, does not have a Fourier transform in either time or space. However, independently, Wiener and Khintchine were able to prove mathematically that the *autocorrelation* of such signal does have a Fourier transform. Furthermore, this function is the *power spectrum* of the random signal. This relationship is known as the Wiener–Khintchine theorem (see chapter X in reference [11]),

$$\int_{-\infty}^{\infty} \int_{V} \Lambda(\boldsymbol{\rho}, \tau) \ e^{i(\omega\tau - \mathbf{k}\cdot\boldsymbol{\rho})} d^3\rho d\tau = S(\mathbf{k}, \omega). \tag{8.23}$$

The inverse relationship also holds,

$$\Lambda(\boldsymbol{\rho}, \tau) = \int_{-\infty}^{\infty} \int_{V} S(\mathbf{k}, \omega) \ e^{-i(\omega\tau - \mathbf{k}\cdot\boldsymbol{\rho})} d^3\rho d\tau. \tag{8.24}$$

In equations (8.23) and (8.24), we maintained our sign convention for the Fourier transform, introduced in volume 1, section 3 [1], whereby the space-time function Λ is obtained as a superposition of monochromatic plane waves, denoted by $e^{-i(\omega\tau - \mathbf{k}\cdot\boldsymbol{\rho})}$.

Note that, by definition, the power spectrum of a stationary signal is a deterministic function with *real* and *positive* values. Because it is integrable, S can be normalized to unit area to represent a *probability density*, $S(\mathbf{k}, \omega)/\int S(\mathbf{k}, \omega)d^3kd\omega$. Its Fourier transform, a normalized version of Λ, is the *characteristic function* associated with a random process. Therefore, up to this normalization constant the autocorrelation function defined by equation (8.24) is nothing more than the *frequency-averaged* monochromatic plane wave associated with the random field, as seen in section 8.1,

$$\langle e^{-i(\omega t - \mathbf{k}\cdot\boldsymbol{\rho})} \rangle_{\mathbf{k}, \omega} \propto \int_{-\infty}^{\infty} \int_{V} S(\mathbf{k}, \omega) \cdot e^{-i(\omega\tau - \mathbf{k}\cdot\boldsymbol{\rho})} d^3\rho d\tau$$
$$= \Lambda(\boldsymbol{\rho}, \tau). \tag{8.25}$$

Thus, the spatiotemporal correlation function has the very interesting physical interpretation of a *monochromatic plane wave*, *averaged* over all spatial and temporal frequencies.

Note that for deterministic signals that have Fourier transforms, the Wiener–Khintchine theorem reduces to the *correlation theorem*, $f \otimes f \leftrightarrow \left| \tilde{f} \right|^2$. This is a general property of the Fourier transform, described in section volume 1, section 4.3 [1]. Therefore, the great importance of the Wiener–Khintchine theorem is due to its applicability to random signals, which do not possess a Fourier transform.

8.2.2 Spatial correlations of monochromatic fields

By taking the Fourier transform of equation (8.24) with respect to time, we obtain what is referred to as the *cross-spectral density* [12]

$$W(\boldsymbol{\rho}, \omega) = \int \Lambda(\boldsymbol{\rho}, \tau)e^{i\omega\tau}d\tau$$
$$= \langle U(\mathbf{r}, \omega)U^*(\mathbf{r} + \boldsymbol{\rho}, \omega) \rangle_{\mathbf{r}}. \tag{8.26}$$

The cross-spectral density function was used previously by Wolf to describe the *second-order statistics* of optical fields, i.e., the Fourier transform of the temporal cross-correlation between two distinct points, $W_{12}(\mathbf{r_1}, \mathbf{r_2}, \omega) = \int \Gamma_{12}(\mathbf{r_1}, \mathbf{r_2}, \tau) \cdot e^{i\omega\tau} d\tau$ [12, 13]. This function describes the similarity in the field fluctuations of two points, like, for example in the two-slit Young interferometer. Note that two points are always fully correlated if the light is monochromatic, because, at most, the field at the two points can differ by a constant phase shift. However, across an entire plane, the phase distribution is a random variable. Therefore, in order to capture the spatial correlations in an *ensemble-averaged* sense, which is most relevant to imaging, we will use the spatially averaged version of $W(\boldsymbol{\rho}, \omega)$, as defined in equation (8.26).

Since $W(\boldsymbol{\rho}, \omega)$ is a spatial correlation function, it can be expressed via a Fourier transform in terms of a *spatiotemporal power spectrum*, $S(\mathbf{k}, \omega)$, as described by the Wiener–Khintchine theorem, which in a 2D plane, reads

$$S(\mathbf{k_{\perp}}, \omega) = \iint W(\boldsymbol{\rho}_{\perp}, \omega) \cdot e^{-i\mathbf{k_{\perp}} \cdot \boldsymbol{\rho}_{\perp}} d^2\boldsymbol{\rho}_{\perp}$$
$$W(\boldsymbol{\rho}_{\perp}, \omega) = \iint S(\mathbf{k_{\perp}}, \omega) \cdot e^{i\mathbf{k_{\perp}} \cdot \boldsymbol{\rho}} d^2\mathbf{k_{\perp}}$$

(8.27)

An important question that arises often in both astronomy and microscopy is: how does the spatial correlation of the field change upon propagation? We will discuss propagation of field correlations much more rigorously later. For now, we seek to acquire an intuitive picture about this interesting phenomenon.

For extended sources that are far away from the detection plane, as in figure 8.3(a), the size of the source may have a smoothing effect on the Fourier transform in equation (8.27). Smoothness over a certain scale indicates that the field is spatially correlated over that scale. For example, for a uniform rectangular source, the Fourier transform is a *sinc* function. Thus, along the x-axis, this correlation distance, x_c, is obtained by writing explicitly the spatial frequency argument of the sinc function,

$$\frac{2\pi}{a} = k_x$$
$$= \frac{2\pi}{\lambda z} x_c.$$

(8.28)

We can conclude that the correlation area of the field generated by the source in the far zone is of the order of (figure 8.3(b))

$$A_C = x_c^2$$
$$= \lambda^2 \left(\frac{z}{a}\right)^2.$$
$$= \frac{\lambda^2}{\Omega}$$

(8.29)

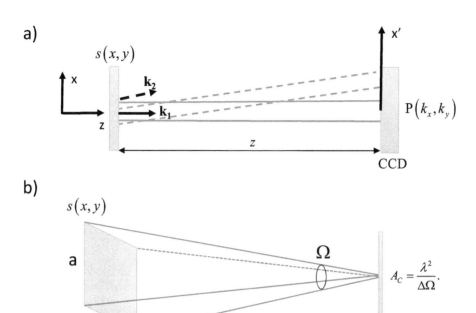

Figure 8.3. (a) Measuring the spatial power spectrum of the field from the source s via Fraunhofer propagation in free space: $\mathbf{k}_{1,2}$ wavevectors, P power spectrum. (b) Coherence area in the far-zone of an extended source is inversely proportional to the solid angle subtended by the source. The solid angle is a measure of spatial bandwidth, namely the spread of the transverse wavevector in the two directions.

Equation (8.29) was expressed in terms of the solid angle Ω subtended by the source, which in the far-zone can be written as, $\Omega = \dfrac{a^2}{z^2}$ (figure 8.3(b)). Note that the solid angle is a measure of the spatial bandwidth, as a/z describes the maximum tilt angle of the wavevector emerging from the source and reaching the plane z. Thus, the maximum transverse wavevector along x is proportional to a/z, $k_x^{max} \propto \beta a/z$, and the solid angle is proportional to the product of the transverse $\mathbf{k}$-vector spreads in the two directions. $\Omega \propto k_x^{max} k_y^{max}$.

8.2.3 Temporal correlations of plane waves

Let us now have the discussion analogous to that in section 8.2.2, where we now investigate the *temporal* correlations of fields at a particular spatial frequency $\mathbf{k}$ (or, equivalently, a plane wave propagating along a certain direction). Taking the *spatial* Fourier transform of Λ in equation (8.19) we obtain the *temporal correlation function* for the plane wave,

$$\Gamma(\mathbf{k}, \tau) = \iint \Lambda(\tilde{\rho}, \tau) e^{-i\mathbf{k}\rho} d^2\mathbf{r}$$
$$= \langle U(\mathbf{k}, t) U^*(\mathbf{k}, t + \tau)\rangle_t$$

(8.30)

The autocorrelation function Γ is relevant in interferometric experiments of the type illustrated in figure 8.4. The temporal correlation Γ is the Fourier transform of the power spectrum,

$$\Gamma(\mathbf{k}, \tau) = \int_{-\infty}^{\infty} S(\mathbf{k}, \omega) \ e^{-i\omega\tau}d\omega$$

$$S(\mathbf{k}, \omega) = \int_{-\infty}^{\infty} \Gamma(\mathbf{k}, \tau) \ e^{i\omega\tau}d\omega.$$

(8.31)

Thus, Γ can be determined via spectroscopic measurements, as exemplified in figure 8.5. By using a grating (a prism, or any other dispersive element), we can

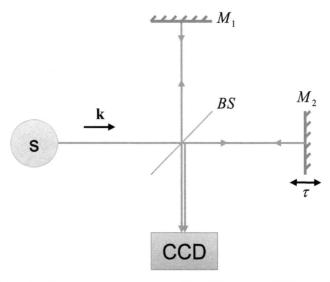

Figure 8.4. Michelson interferometry: s source, $\mathbf{k}$ wavevector, $M_{1,2}$ mirrors, BS beam splitter, τ temporal delay introduced by displacing mirror M_2.

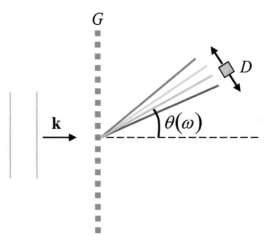

Figure 8.5. Spectroscopic measurement using a grating: G grating, D detector, θ diffraction angle.

'disperse' different colors at different angles, such that a rotating detector (*goniometer*) can measure $S(\omega)$ directly. In order to estimate the coherence time for a broad band field, let us assume a Gaussian spectrum centered at frequency ω_0, and having the standard deviation width $\Delta\omega$,

$$S(\omega) = S_0 e^{-\left(\frac{\omega-\omega_0}{\sqrt{2}\,\Delta\omega}\right)^2}, \tag{8.32}$$

where S_0 is a constant. Using the Fourier transform properties of a Gaussian function (volume 1, chapter 4 [1]), the autocorrelation function is also a Gaussian, modulated by a sinusoidal function, as a result of the Fourier shift theorem

$$\Gamma(\tau) = \Gamma_0 e^{-\left(\frac{\Delta\omega\tau}{\sqrt{2}}\right)^2} e^{i\omega_0\tau} \tag{8.33}$$

From equation (8.33), we see that if we define the width of Γ as the *coherence time*, we obtain

$$\tau_c \propto \frac{1}{\Delta\omega}, \tag{8.34}$$

and the coherence length

$$l_c = c\tau_C$$
$$\propto \frac{\lambda^2}{\Delta\lambda} \tag{8.35}$$

The coherence length depends on the spectral bandwidth in an analogous fashion to the coherence area dependence on solid angle (equation (8.29)). This is not surprising as both types of correlation depend on their respective frequency bandwidth.

8.3 Coherence mode decomposition of random fields

Recently, it has been shown that a statistically homogeneous field can always be decomposed into a plane wave basis and, furthermore, its second-order statistics is fully captured by a deterministic signal associated with the random field [14]. In 1982, Wolf introduced the coherence mode decomposition, which showed that the cross-spectral density, $W(\mathbf{r}_1, \mathbf{r}_2; \omega) = \langle U^*(\mathbf{r}_1; \omega)U(\mathbf{r}_2; \omega)\rangle$, of a stochastic field, U, can be represented in the form of Mercer's expansion [12],

$$W(\mathbf{r}_1, \mathbf{r}_2; \omega) = \sum_n \lambda_n(\omega)\phi_n^*(\mathbf{r}_1; \omega)\phi_n(\mathbf{r}_2; \omega), \tag{8.36}$$

where the sum is uniform and absolute convergent. The angular bracket denotes the ensemble average over the spatially varying ensemble. The functions ϕ_n are referred to as coherent modes. They are orthogonal eigenfunctions of W corresponding to the eigenvalues λ_n, as defined by the Fredholm integral equation

$$\int_D W(\mathbf{r}_1, \mathbf{r}_2; \omega)\phi_n(\mathbf{r}_1; \omega)d^3\mathbf{r}_1 = \lambda_n(\omega)\phi_n(\mathbf{r}_2; \omega). \tag{8.37}$$

In equation (8.37), D is the 3D spatial domain of interest, e.g., for a primary source, D is a volume containing the source. Because W is Hermitian, all the eigenvalues λ_n are non-negative.

Let us consider a *statistically homogeneous* field at a certain plane of interest, where our coordinate is $r_{1,2\perp} \in D$, $D \subset \mathbf{R}^2$. The cross-spectral density simplifies to $W(r_{1\perp}, r_{2\perp}; \omega) = W(r_{2\perp} - r_{1\perp}; \omega)$ and equation (8.37) becomes

$$\left[W \circledcirc \phi_n\right](r_\perp, \omega) = \lambda_n(\omega)\phi_n(r_\perp, \omega), \tag{8.38}$$

where $\circledcirc$ denotes the two dimensional convolution over $r_\perp$. It can be shown that a set of plane waves, $\phi_n(r_\perp; \omega) = e^{ik_{n\perp}(\omega)\cdot r_\perp}$, satisfies equation (8.38) [14]. Using the property of the convolution with a complex exponential, we see immediately that $|W \circledcirc e^{ik_{n\perp}(\omega)\cdot r_\perp}|(r_\perp; \omega) = S(k_{n\perp}; \omega)e^{ik_{n\perp}(\omega)\cdot r_\perp}$ with S the Fourier transform of W. As a result, equation (8.38) yields the following eigenvalues

$$\lambda_n(\omega) = S(k_{n\perp}; \omega). \tag{8.39}$$

Clearly, these eigenvalues are non-negative because S is a (spatiotemporal) power spectrum. Furthermore, it has been shown that the plane wave decomposition of U is unique in the non-degenerate case, i.e., when the power spectrum, $S(k_\perp; \omega)$, is injective in the $k_\perp$-space [15].

8.4 Deterministic signal associated with a random stationary field

Notwithstanding uniqueness, we can always expand a statistically homogeneous field, U, in a Karhunen–Loéve [13] series using plane waves as the coherent modes, $U(r_\perp; \omega) = \sum_n a(k_{\perp n}; \omega)e^{ik_{\perp n}\cdot r_\perp}$, where $a(k_{\perp n}; \omega)$ are random coefficients characterizing the ensemble of the spectral component $U(k_{\perp n}; \omega)$. It is worth clarifying the nature of this randomness. The coefficients $a(k_{\perp n}; \omega)$ do not depend on the time variable, t. The distribution of these coefficients is described by an underlying probability density for mode $k_{\perp n}$. For example, we can define an average, variance, etc, associated with the $k_{\perp n}$ distribution of $a(k_{\perp n}; \omega)$. As emphasized early on by Wolf, the vast majority of fields encountered in practice are ergodic (see p 345, in reference [12]). This allows us to replace the ensemble averages by averages over the measurable domain, i.e., space or time. Here, we employ *spatial ergodicity* to express W in terms of a spatial average, namely

$$\begin{aligned} W(\Delta r_\perp; \omega) &= \langle U^*(r_{1\perp}; \omega)U(r_{1\perp} + \Delta r_\perp; \omega)\rangle_{r_{11}} \\ &= \sum_m \sum_n a^*(k_{\perp m}; \omega)a(k_{\perp n}; \omega)e^{ik_{\perp m}\cdot\Delta r_\perp}\int_D d^2r_{11}e^{i(k_{\perp m}-k_{\perp n})\cdot r_{11}} \\ &= \sum_m \sum_n a^*(k_{\perp m}; \omega)a(k_{\perp n}; \omega)e^{ik_{\perp m}\cdot\Delta r_\perp}\delta^{(2)}(k_{\perp m} - k_{\perp n}) \\ &= \sum_n | a(k_{\perp n}; \omega)|^2 e^{ik_{\perp n}\cdot\Delta r_\perp}. \end{aligned} \tag{8.40}$$

Thus, there areno random quantities left in equation (8.40), i.e., for statistically homogeneous fields, the cross-spectral density and, implicitly, the power spectrum, are deterministic functions. Therefore, taking the Fourier transform of equation (8.40), we find $| a(\mathbf{k}_{\perp n}; \omega)|^2 = S(\mathbf{k}_{\perp n}; \omega)$, which is the spatiotemporal power spectrum of the field. We conclude that spatially ergodic fields can be decomposed into plane waves, whose amplitudes are deterministic and given by the power spectrum, $S(\mathbf{k}_{\perp n}; \omega)$.

On closer inspection, we see that the spatial *randomness* is only due to the *spectral phase,* which does not factor into the spatial correlation, W, or power spectrum, S. Thus, spatially ergodic fields can be expressed quite generally as

$$U(\mathbf{r}_\perp; \omega) = \sum_n \sqrt{S(\mathbf{k}_{\perp n}; \omega)}\, e^{i[\mathbf{k}_{\perp n} \cdot \mathbf{r}_\perp + \varphi(\mathbf{k}_{\perp n}; \omega)]}. \qquad (8.41)$$

In equation (8.41), $\varphi(\mathbf{k}_{\perp n}; \omega)$ is the spectral phase, the only random variable in the expansion. Note that the values of $\varphi(\mathbf{k}_{\perp n}; \omega)$ do not affect the cross-spectral density, W, associated with U. Therefore, we can construct a *fictitious signal,* referred to as the *deterministic signal associated with a random field* [14], V, by taking $\varphi(\mathbf{k}_{\perp n}; \omega) = 0$, i.e., $V(\mathbf{r}_\perp; \omega) = \sum_n \sqrt{S(\mathbf{k}_{\perp n}; \omega)}\, e^{i\mathbf{k}_{\perp n} \cdot \mathbf{r}_\perp}$, which has the same spatial correlation, as the random field U, satisfying $\int D V^*(\mathbf{r}_{\perp 1}; \omega) V(\mathbf{r}_{\perp 1} + \Delta \mathbf{r}_\perp; \omega) d^2\mathbf{r}_{\perp 1} = W(\Delta \mathbf{r}_\perp; \omega)$. In problems related to propagation of correlations, it is much simpler to propagate the deterministic field to the plane of interest and then calculate its correlation, rather than propagating the correlation function itself (figure 8.6).

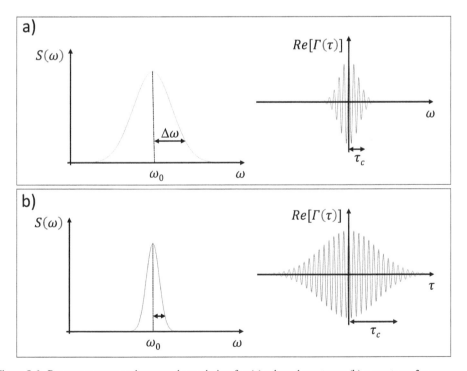

Figure 8.6. Power spectrum and temporal correlation for (a) a broad spectrum; (b) a spectrum 3× narrower. The central frequency, ω_0, is the same.

We illustrate the relevance of the results presented here with two problems of spatial and temporal field distributions of common interest: coherent versus incoherent illumination in a microscope (figure 8.7) and, respectively, pulsed versus CW emission of a broadband laser (figure 8.8). In microscopy, the spatial coherence of the illumination is governed by the numerical aperture of the condenser lens, $NA_c = \sin(\theta_{max})$, assuming free space with refractive index of $n=1$ (figure 8.7(a) and (b)). Each point in the condenser aperture generates one mode of the illuminating field, $|a(\mathbf{k}_{\perp n}; \omega)| \exp[i\mathbf{k}_{\perp n}\cdot \mathbf{r}_\perp + \varphi(\mathbf{k}_{\perp n}; \omega)]$. The spectral phase distribution, $\varphi(\mathbf{k}_{\perp n}; \omega)$, depends on the nature of the extended primary or secondary source (diffuser in the case of a microscope).

Figure 8.8(a) illustrates how the coefficients $a(\mathbf{k}_{\perp n}; \omega)$, are generated by an extended source in microscopy. Note each wavevector, $\mathbf{k}_{\perp n}$ corresponds to a different location on the extended source. The numerical aperture of the condenser determines how many components $\mathbf{k}_{\perp n}$ contribute to the total field $U(\mathbf{r}_\perp; \omega)$. An intensity measurement over the area of this extended source gives the spatial power spectrum, $S(\mathbf{k}_{\perp n}; \omega)$ (figure 8.7(c)). The 2D Fourier transform of the spatial power spectrum gives the cross-spectral density $W(\mathbf{r}_{1\perp} - \mathbf{r}_{2\perp}; \omega)$, as shown in figure 8.7(d). As pointed

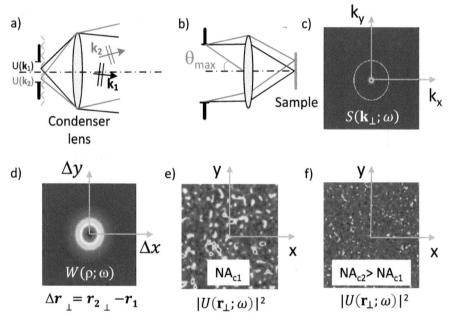

Figure 8.7. The condenser controls the spatial coherence of the illumination field in a microscope. (a) The spatial coherence of the illumination is governed by the size of the condenser aperture. Each point in the aperture generates one mode for the illuminating field $(\mathbf{k}_\perp; \omega)\exp[i(\mathbf{k}_\perp \cdot \mathbf{r}_\perp)]$. The phase of each mode, $\arg[U(\mathbf{k}_\perp; \omega)]$, depends on the nature of the extended source. (b) Conversely, each plane wave mode from the illumination aperture focuses into a small point at the sample plane. The numerical aperture of the condenser is defined as $NA_c = \sin(\theta_{max})$. (c) Spatial power spectrum, i.e., the irradiance distribution at the condenser aperture plane. The yellow circle denotes the diffraction limit, the entrance pupil of the objective. (d) Cross-spectral density, the Fourier transform of S. (e) Irradiance distribution, i.e., speckle field, at the sample plane. (f) Speckle field for NA_c larger than in (e), such that the speckle size is smaller.

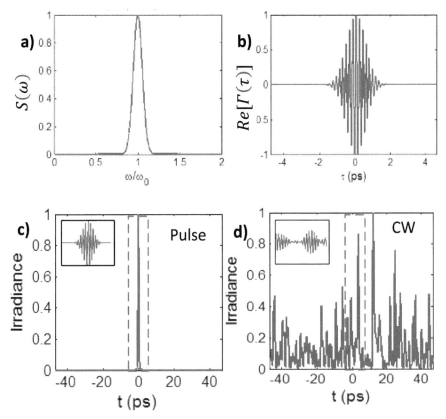

Figure 8.8. (a) Gaussian power spectrum of a low-coherence source at central wavelength $\lambda_o = 545$ nm, FWHH = 30 nm. (b) The temporal correlation function of the same source in (a) obtained by Fourier transforming the power spectrum. Envelope of this function is controlled by the bandwidth of the source. Larger bandwidth creates shorter envelope and vice-versa. The oscillation inside the envelope is given by the central optical frequency ω_o. (c) Irradiance of a short pulse with the same correlation function as in (b) generated by setting the phase value of 0 to the spectral component. The inset shows the real part of the pulse around the region $t = 0$ ps inside the dashed box. (d) Irradiance of a CW laser with the same correlation function in (b) with uniformly distributed spectral phase in $[-\pi/\pi]$. The inset displays the real part of the continuous wave for the region specified by the dashed box.

out above, the knowledge of the spatial power spectrum $S(\mathbf{k}_{\perp n}; \omega)$ does not fully determine the coefficients $a(\mathbf{k}_{\perp n}; \omega)$ but only their amplitudes, $| a(\mathbf{k}_{\perp n}; \omega) |$. The phase of these coefficients defines the field distribution. Setting the phase of these coefficients to zero brings a tightly focused field $U(\mathbf{r}_{\perp}; \omega)$. Meanwhile, the uniformly distributed phase from $(-\pi, \pi)$ generates the speckle pattern, although the fields in these two cases have the same cross-spectral density $W(\mathbf{r}_{1\perp} - \mathbf{r}_{2\perp}; \omega)$. Note that the waist of the focused beam or the granularity of the speckle pattern ('speckle size') depend solely on the spatial power spectrum $\widetilde{W}(\mathbf{k}_{\perp n}; \omega)$, which is itself controlled by the numerical aperture of the illumination (figure 8.7(e–f)). A smaller value of this numerical aperture gives a larger coherence area and a larger speckle size.

These arguments can be translated in full to temporally random fields. Assuming a plane wave of wavevector $\mathbf{k}$ (spatially deterministic), we can show at once that, temporally, the field can be decomposed into monochromatic waves with random phases, namely

$$U(\mathbf{k}, t) = \sum_n \sqrt{S(\mathbf{k}, \omega_n)}\, e^{i[\omega_n t + \varphi(\mathbf{k}, \omega_n)]}. \tag{8.42}$$

If S is an injective function on ω, the monochromatic wave decomposition is unique. Figure 8.8 illustrates how the difference between a CW source and a light pulse of the same power spectrum is entirely in the distribution of the spectral phase $\varphi(\omega_n)$. We assume a Gaussian power spectrum of central frequency ω_o (figure 8.8(a)). The corresponding temporal correlation function, $\Gamma(\tau)$, of this source is shown in figure 8.8(b). Figure 8.8(c) shows the irradiance of the field distribution when the spectral phase distribution is set to zero. When the spectral phase is randomly uniformly distributed over $(-\pi, \pi)$, while keeping the spectral amplitudes fixed, we obtain CW operation figure 8.8(d). Both signals have the same correlation function as shown in figure 8.8(b). Therefore, not surprisingly, one method for obtaining short pulses, e.g., in Ti: Saph lasers, is referred to as *mode-locking*, which means converting the spectral phase from a random variable, $\{\varphi(\omega)\}$, to a deterministic function, $\varphi(\omega)$.

8.5 Propagation of field correlations: intuitive picture

Coherence properties of light can change significantly upon field propagation. A testament to this reality is the fact that, for a long time, interferometric experiments have been performed using light from stars, such as the Sun, which are *incoherent sources*. Before we go into a quantitative description of how propagation influences spatial and temporal correlations, it is instructive to consider the following intuitive picture (see, e.g., Mandell and Wolf, chapter 4).

Let us consider two uncorrelated sources, A and B, emitting random fields, U_A and U_B (figure 8.9). For simplicity, let us consider them as point sources. The complete lack of mutual correlation in time can be described by a vanishing temporal cross-correlation function,

$$\Gamma_{AB}(\tau) = \langle U_A(t) U_B \times (t + \tau) \rangle = 0, \tag{8.43}$$

where $\langle \ \rangle$ indicates ensemble averaging. However, we assume that the individual temporal *autocorrelations* are non-vanishing, at least for some time interval, $|\tau| < \tau_c$,

$$\begin{aligned} \Gamma_{A,\,B}(\tau) &= \langle U_{A,\,B}(t) U_{A,\,B} \times (t + \tau) \rangle \\ |\Gamma_{A,\,B}(\tau)| &\neq 0, |\tau| < \tau_c \end{aligned} \tag{8.44}$$

Having established that the two sources are completely uncorrelated, next we ask the question: are the fields observed at two points (A' and B') correlated? The crucial aspect of the thought experiment is that each observation point receives light from both sources. Thus, the fields at the two points are a linear combination of U_A and U_B

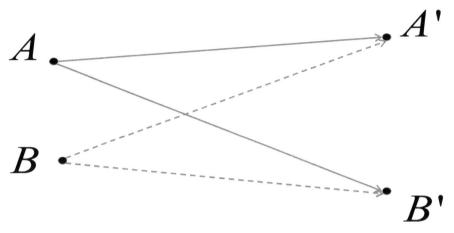

Figure 8.9. Intuitive picture for understanding the effects of propagation on coherence: A and B are uncorrelated sources, A' and B' are observation points, which register some degree of correlation in their signals.

$$U_{A'} = a_1 U_A + b_1 U_B$$
$$U_{B'} = a_2 U_A + b_2 U_B. \tag{8.45}$$

The coefficients $a_{1,2}$, $b_{1,2}$ are generally complex and contain information about the free-space propagation from the sources to the observation points.

In order to describe the field correlation at the two observation points, we write down the cross-correlation function,

$$\langle U_{A'}(t) U_{B'}(t + \tau)\rangle$$
$$= \langle [a_1 U_A(t) + b_1 U_B(t)][a_2 \times U_A \times (t + \tau) + b_2 \times U_B \times (t + \tau)]\rangle \tag{8.46}$$
$$= a_1 a_2 \times \langle U_A(t) U_A \times (t + \tau)\rangle + b_1 b_2 \times \langle U_B(t) U_B \times (t + \tau)\rangle + 0 + 0.$$

Note that the *cross terms*, of the form in equation (8.43), vanish because we assumed independent sources. However the *autocorrelation* (diagonal) terms survive the averaging, i.e., the fields at A' and B' are correlated at least within some temporal interval, $|\tau| < \tau_c$. This is an important result that shows how one can detect interference fringes at a certain distance from uncorrelated sources. In essence, this result indicates that two linear combinations (*superpositions*) of *uncorrelated* signals can exhibit correlations. This simple example highlights that the temporal and spatial correlations are interconnected.

8.6 Stochastic wave equation

In order to understand quantitatively how field correlations propgate, here we generalize the result in the previous section by studying the propagation of field correlations from an arbitrary source s that emits a random field. We start with the scalar wave equation that has as driving term in the form of a random source,

$$\nabla^2 U(\mathbf{r}, t) - \frac{1}{c^2}\frac{\partial^2 U(\mathbf{r}, t)}{\partial t^2} = s(\mathbf{r}, t). \tag{8.47}$$

The source signal, s, as introduced in volume 2 (chapter 15) [4] and figure 8.10 can be regarded as a realization of the fluctuating source field (U is the complex analytic signal associated with the real propagating field). For generality, here we consider a 3D spatial field distribution, $\mathbf{r} = (x, y, z)$.

Because equation (8.47) has a random (stochastic) driving term, it is referred to as a *stochastic differential equation*. Notoriously, Langevin introduced such an equation (the *Langevin equation*) to describe Brownian motion of particles [16]. The key difference with respect to the *deterministic* wave equation used in all other chapters of this book, is that the field s does not have a prescribed form, i.e., we cannot express the source field via an analytic function. Instead, it is known only through average quantities, e.g., the autocorrelation function or, equivalently, the power spectrum, as defined in volume 2 (chapter 15) [4]. For simplicity, we assume the source field to be stationary (at least in the wide sense) and statistically homogeneous.

Particular types of sources that are commonly encountered in practice include
(i) the *white noise source*, for which the spatiotemporal autocorrelation function reads

$$\Lambda(\boldsymbol{\rho}, \tau) = \langle s(\mathbf{r}, t)s \times (\mathbf{r} + \boldsymbol{\rho}, t + \tau)\rangle$$
$$= \text{const.}\delta(\tau)\delta^{(2)}(\boldsymbol{\rho}) \tag{8.48}$$
$$S(\mathbf{k}, \omega) = \text{const.}$$

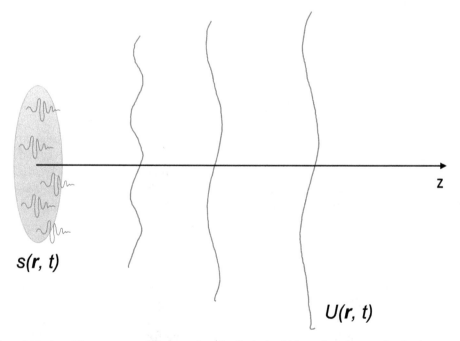

Figure 8.10. An arbitrary source emitting a random signal, $s(\mathbf{r}, t)$, which consists of uncorrelated point sources.

(ii) natural linewidth (Lorentzian temporal spectrum) source

$$\Lambda(\boldsymbol{\rho}, \tau) = f(\boldsymbol{\rho})e^{-\frac{|\tau|}{\Delta\tau}}$$

$$S(\mathbf{k}, \omega) = \widetilde{f}(\mathbf{k})\frac{1}{1 + (\omega\Delta\tau)^2}. \tag{8.49}$$

(iii) Gaussian sources

$$\Lambda(\boldsymbol{\rho}, \tau) = \text{const}\, e^{-\frac{\tau^2}{2\Delta\tau^2}} \cdot e^{-\frac{\rho^2}{2\Delta\rho^2}} \tag{8.50}$$

$$S(\mathbf{k}, \omega) = \text{const}\, e^{-\frac{(\omega\Delta\tau)^2}{2}} \cdot e^{-\frac{(k\Delta\rho)^2}{2}}$$

These three types of source statistics describe various phenomena: (i) denotes complete lack of correlations in both time and space, (ii) a Lorentzian (temporal) spectrum is associated with the emission of an isolated two-level atomic system (recall section 7.3, volume 2 [4]), and (iii) describes a Gaussian spectrum that is typically due to the superposition of many independent events and the central limit theorem (e.g., recall section 7.3 for Doppler broadening of laser linewidths).

For a source at thermal equilibrium, we found the spectral density given by Planck's law (recall volume 2, chapter 6 [4]),

$$S(\omega) = \text{const}\frac{\omega^3}{e^{\frac{\hbar\omega}{k_B T}} - 1}, \tag{8.51}$$

where $\hbar$ is Planck's constant $h/2\pi$, k_B is Boltzmann's constant and T is the absolute temperature. Thus, for studying the field propagating from a thermal source, the temporal component of the source fluctuations is governed by equation (8.51).

Regardless of the particular physical nature of the source and, thus, of the spatiotemporal autocorrelation function that describes its fluctuations, we use the stochastic wave equation (equation (8.47)) to solve for the *autocorrelation of U* and *not U* itself. In order to achieve this, we take the spatiotemporal autocorrelation of equation (8.47) on both sides,

$$\left\langle \left[\nabla_1^2 U(\mathbf{r}, t) - \frac{1}{c^2}\frac{\partial^2 U(\mathbf{r}, t)}{\partial t^2} \right] \right.$$
$$\left. \left[\nabla_2^2 U(\mathbf{r} + \boldsymbol{\rho}, t + \tau) - \frac{1}{c^2}\frac{\partial^2 U(\mathbf{r} + \boldsymbol{\rho}, t + \tau)}{\partial(t + \tau)^2} \right]^* \right\rangle \tag{8.52}$$
$$= \langle s(\mathbf{r}, t)s \times (\mathbf{r} + \boldsymbol{\rho}, t + \tau)\rangle$$
$$= \Lambda_s(\boldsymbol{\rho}, \tau),$$

where the angle brackets indicate ensemble averaging, ∇_1^2 is the Laplacian with respect to coordinate $\mathbf{r}$, ∇_2^2 with respect to coordinate $\mathbf{r} + \boldsymbol{\rho}$, and Λ_s is the spatiotemporal autocorrelation function of s. Since we assumed wide sense stationarity and statistical homogeneity, which gives Λ_s dependence only on the differences ρ and τ, all the derivatives in equation (8.52) can be taken with respect to the shifts (see p 194 in Mandel and Wolf), namely

$$\nabla_1^2 = \nabla_2^2 = \frac{\partial^2}{\partial \rho_x^2} + \frac{\partial^2}{\partial \rho_y^2} + \frac{\partial^2}{\partial \rho_z^2}$$

$$\frac{\partial}{\partial t^2} = \frac{\partial}{\partial (t + \tau)^2} = \frac{\partial}{\partial \tau^2}. \tag{8.53}$$

After these simplifications, equation (8.52) can be rewritten as

$$\left(\nabla^2 - \frac{1}{c^2} \frac{\partial}{\partial \tau^2} \right) \left(\nabla^2 - \frac{1}{c^2} \frac{\partial}{\partial \tau^2} \right) \Lambda_U(\boldsymbol{\rho}, \tau) = \Lambda_s(\boldsymbol{\rho}, \tau) \tag{8.54}$$

where Λ_U is the spatiotemporal autocorrelation of U, $\Lambda_U(\boldsymbol{\rho}, \tau) = \langle U(\mathbf{r}, t) U \times (\mathbf{r} + \boldsymbol{\rho}, t + \tau) \rangle$.

Equation (8.54) is a differential equation that relates the autocorrelation of the propagating field, U, with that of the source, s. Due to the Wiener–Khintchine theorem, we know that both Λ_U and Λ_s have Fourier transforms, which are their respective spectra, S_U and S_s, respectively. Therefore, we can solve this differential equation, as usual, by Fourier transforming it with respect to both ρ and τ,

$$\left(\beta_0^2 - k^2 \right) \left(\beta_0^2 - k^2 \right) S_U(\mathbf{k}, \omega) = S_s(\mathbf{k}, \omega) \tag{8.55a}$$

$$S_U(\mathbf{k}, \omega) = \frac{S_s(\mathbf{k}, \omega)}{\left(\beta_0^2 - k^2 \right)^2} \tag{8.55b}$$

In equation (8.55), we used the differentiation property of the Fourier transform, $\nabla \rightarrow i\mathbf{k}$, $\frac{\partial}{\partial \tau} \rightarrow -i\omega$. Equation (8.55b) represents the full solution of equation (8.55(a)) in the $\mathbf{k} - \omega$ representation; it gives an expression for the spectrum of the propagating field, S_U, with respect to the spectrum of the source, S_s. Note that here the function $(\beta_0^2 - k^2)^{-2}$ is a filter function (*transfer function*), which incorporates all the effects of free space propagation. Note that, in solving the deterministic wave equation, we arrived at $(\beta_0^2 - k^2)^{-1}$ as a filter function for the field propagation (section 4.3).

8.7 Wave equation for the deterministic signal associated with a random field

It is quite remarkable that the second-order statistics of a fluctuating field is contained in its power spectrum $S(\mathbf{k}, \omega)$, a *real*, *positive* function. The assumed wide sense stationarity ensures that the spectrum does not change in time; it is a deterministic function. Therefore, we can mathematically introduce a *spectral amplitude*, $V(\mathbf{k}, \omega)$, via a square root operation,

$$V(\mathbf{k}, \omega) = \sqrt{S(\mathbf{k}, \omega)}, \tag{8.56}$$

which contains full information about the field fluctuations. Of course, V has a Fourier transform, provided it is *modulus integrable*. The fact that $V(\mathbf{k}, \omega)$ is *modulus-squared integrable* (the spectrum contains finite energy) does not ensure that $\int |V| d\omega d\mathbf{k} < \infty$. However, for most spectral distributions of interest, S decays fast enough at infinity, such that its square root is integrable as well.

Therefore, we can introduce a *deterministic signal* associated with the *random field* as the Fourier transform inverse of $V(\mathbf{k}, \omega)$,

$$V(\mathbf{r}, t) = \int_V \int_{-\infty}^{\infty} V(\mathbf{k}, \omega) e^{-i(\omega t - \mathbf{k} \cdot \mathbf{r})} d\omega d^3\mathbf{k}$$

$$V(\mathbf{k}, \omega) = \int_{V_k} \int_{-\infty}^{\infty} V(\mathbf{r}, t) e^{i(\omega t - \mathbf{k} \cdot \mathbf{r})} dt d^3\mathbf{r}.$$

(8.57)

where V_k is the 3D domain of the spatial frequency. Thus, taking the square root of equation (8.55b), we can write

$$V_U(\mathbf{k}, \omega) = \frac{V_s(\mathbf{k}, \omega)}{\beta_0^2 - k^2},$$

(8.58)

where $V_U(\mathbf{k}, \omega)$ and $V_s(\mathbf{k}, \omega)$ are the Fourier transforms of the deterministic signals associated with the (random) source and propagating field, respectively. Going to the space-time domain, equation (8.58) indicates that $V_U(\mathbf{r}, t)$ and $V_s(\mathbf{r}, t)$ satisfy the *deterministic* wave equation, i.e.,

$$\nabla^2 V_U(\mathbf{r}, t) - \frac{1}{c^2} \frac{\partial^2 V_U(\mathbf{r}, t)}{\partial t^2} = V_s(\mathbf{r}, t).$$

(8.59)

Comparing our original, *stochastic* wave equation (equation (8.47)) with the *deterministic* version in equation (8.59), it is clear that the only difference is replacing the source field with its *deterministic signal*, which in turn requires that we replace the stochastic propagating field with its deterministic counterpart.

In essence, by introducing the 'fictitious' deterministic signal, we reduced the problem of propagating correlations to solving the common (second-order) wave equation. We can now apply the methods presented in chapter 4 to solve for V_U in 1D, 2D, and 3D, provided we know the spectrum of the source S, which in turn gives $V_s(\mathbf{k}, \omega)$ and $V_s(\mathbf{r}, t)$. Importantly, the solution to the problem must be presented in terms of the *autocorrelation* of $V_U(\mathbf{r}, t)$, $\Lambda_U(\boldsymbol{\rho}, \tau)$, or its *spectrum* $|V_U(\mathbf{k}, \omega)|^2$, as V_U itself exists only as a mathematical construct. Next, we are going to apply this concept to propagating spatial correlations in free space.

8.8 Propagation of spatial coherence: van Cittert–Zernike theorem

One important result in coherence theory is due to van Cittert [17] and Zernike [18]. This *van Cittert–Zernike theorem* establishes the spatial autocorrelation of the field in the *far-zone* of a completely incoherent source (figure 8.10). The result was originally formulated in terms of the *mutual intensity*, defined as

$$J(\mathbf{r}_1, \mathbf{r}_2) = \langle U(\mathbf{r}_1, t)U \times (\mathbf{r}_2, t) \rangle, \tag{8.60}$$

where the angle brackets indicate ensemble averaging over a certain *area* of interest (e.g., we are interested in the field distribution in a plane). This function describes the time-averaged *spatial* similarity (autocorrelation) of the field, and it has been used commonly in statistical optics (see, e.g., reference [19]). The van Cittert–Zernike theorem establishes a relationship between J at the source plane and that of the field in the far zone. Such propagation of correlations has been described in detail by Mandel and Wolf [13]. Here, we derive the main result using the concept of the *deterministic signal* associated with a random field, as follows.

From the basic principles, we anticipate that J depends on all optical frequencies, i.e., it can be expressed as an integral over the frequency domain. Therefore, we choose to work in the frequency domain, which is a more general treatment, keeping in mind that we can always integrate the result over frequencies and obtain the mutual intensity. For simplicity, we assume again wide sense *statistically homogeneous* and stationary fields, such that equation (8.60) simplifies to

$$J(\boldsymbol{\rho}) = \langle U(\mathbf{r}, t)U \times (\mathbf{r} + \boldsymbol{\rho}, t) \rangle. \tag{8.61}$$

Next, we note that this *mutual intensity J* is the *spatiotemporal correlation function* introduced in section 8.2, evaluated at time delay $\tau = 0$,

$$J(\boldsymbol{\rho}) = \Lambda(\boldsymbol{\rho}, \tau = 0). \tag{8.62}$$

It follows that, due to the *central ordinate theorem*, the autocorrelation function evaluated at $\tau = 0$ equals the power spectrum integrated over all frequencies,

$$J(\boldsymbol{\rho}) = \Lambda(\boldsymbol{\rho}, \tau = 0)$$
$$= \int_{-\infty}^{\infty} \Lambda(\boldsymbol{\rho}, \omega)d\omega. \tag{8.63}$$

Therefore, one way to obtain $J(\boldsymbol{\rho})$, perhaps the most convenient, is via the spatiotemporal power spectrum, $S(\mathbf{k}, \omega)$, followed by Fourier transforming with respect to $\mathbf{k}$ and integrating over ω.

In order to derive an expression for the *coherence area* associated with a field propagating from an arbitrary 2D source, we start with the wave equation in terms of the *deterministic signals* associated with the random fields (recall equation (8.59))

$$\nabla^2 V_U(\mathbf{r}, t) - \frac{1}{c^2}\frac{\partial^2 V_U(\mathbf{r}, t)}{\partial t^2} = s_0 V_s(x, y, t)\delta(z), \tag{8.64}$$

where V_U and V_s are the *deterministic signals* associated with the actual signals U and S, respectively, i.e., $|V_U(k, \omega)|^2 = S_U$ and $|V_S(k, \omega)|^2 = S_s$; S is the source field and U the propagating field. We assume a planar source, i.e., infinitely thin along z, described in equation (8.64) by $\delta(z)$. Fourier transforming equation (8.64), we readily obtain the solution in the $\mathbf{k} - \omega$ domain

$$V_U(\mathbf{k}, \omega) = \frac{V_S(\mathbf{k}_\perp, \omega)}{\beta_0^2 - k^2}, \tag{8.65}$$

where $\mathbf{k}_\perp = (k_x, k_y)$ and $k^2 = k_x{}^2 + k_y{}^2 + k_z{}^2$, $\beta_0 = \omega/c$. Next, we represent the propagating field in terms of $(\mathbf{k}_\perp, z)$. Thus, using the partial fraction decomposition,

$$\frac{1}{\beta_0^2 - k^2} = \frac{1}{\gamma^2 - k_z^2}$$
$$= \frac{1}{2\gamma}\left[\frac{1}{\gamma - k_z} + \frac{1}{\gamma + k_z}\right], \tag{8.66}$$

where $\gamma = \sqrt{\beta_0 - k_x^2 - k_y^2}$. Eliminating the negative frequency (*inward*) term, $1/(\gamma + k_z)$, we arrive at

$$V_U(\mathbf{k}, \omega) = \frac{V_S(\mathbf{k}_\perp, \omega)}{2\gamma(\gamma - k_z)}. \tag{8.67}$$

By Fourier transforming with respect to k_z, we obtain the field V_U as function of $\mathbf{k}_\perp$ and z, which is known as the *plane wave decomposition* or *Weyl's formula* (see section 5.2),

$$V_U(\mathbf{k}_\perp, z, \omega) = -iV_S(\mathbf{k}_\perp, \omega)\frac{e^{i\gamma z}}{2\gamma}. \tag{8.68}$$

The magnitude squared on both sides in equation (8.68) yields a z-independent relation in terms of the respective power spectra, $S_U(\mathbf{k}_\perp, \omega) = |V_U(k_z, z, \omega)|^2$ and $S_s(\mathbf{k}_\perp, \omega) = |V_s(\mathbf{k}_\perp, z, \omega)|^2$,

$$S_U(\mathbf{k}_\perp, \omega) = \frac{1}{4}\frac{S_s(\mathbf{k}_\perp, \omega)}{(\beta_0^2 - k_\perp^2)}. \tag{8.69}$$

Equation (8.69) describes the statistical properties of the field at a plane z, assuming that the field is stationary. The coherence area of the field at that plane is inversely proportional to spatial bandwidth of $S_U(\mathbf{k}_\perp, \omega)$. Thus, we can use the variance of the $S_U(\mathbf{k}_\perp, \omega)$ distribution over $\mathbf{k}_\perp$ to compute the coherence area, $A_c = 1/\langle k_\perp^2 \rangle$.

If we assume that the spectrum of the observed field is centered at the origin, i.e., $\langle \mathbf{k}_\perp \rangle = 0$, and isotropic, i.e., depends only on the magnitude of $\mathbf{k}_\perp$ and not its direction, the variance can be simply calculated as the second moment of $k_\perp = |\mathbf{k}_\perp|$

$$\langle k_\perp^2(\omega) \rangle = \frac{\displaystyle\int_{A_{k_\perp}} k_\perp^2 S_U(\mathbf{k}_\perp, \omega)d^2\mathbf{k}_\perp}{\displaystyle\int_{A_{k_\perp}} S_U(\mathbf{k}_\perp, \omega)d^2\mathbf{k}_\perp}, \tag{8.70}$$

where $A_{k_\perp}$ is the $\mathbf{k}_\perp$ domain of integration. Using equation (8.69) to evaluate equation (8.70), the variance is obtained at once

$$\langle k_\perp^2(\omega) \rangle = \beta_0^2 - \frac{\displaystyle\int_{A_{\mathbf{k}_\perp}} S_s(\mathbf{k}_\perp, \omega) d^2\mathbf{k}_\perp}{\displaystyle\int_{A_{\mathbf{k}_\perp}} \frac{S_s(\mathbf{k}_\perp, \omega)}{\beta_0^2 - k_\perp^2} d^2\mathbf{k}_\perp}. \tag{8.71}$$

We consider the source as *fully* spatially incoherent at all frequencies ω, i.e., $S_s(\mathbf{k}_\perp, \omega) = S_s(\omega)$, which assumes a delta-function spatial correlation. Full spatial incoherence simplifies equation (8.71) to $\langle k_\perp^2(\omega) \rangle = \beta_0^2 - \int_{A_{k_\perp}} d^2 k_\perp \Big/ \int_{A_{k_\perp}} \big[1 \big/ (\beta_0^2 - k_\perp^2) d^2 k_\perp \big]$. Furthermore, if we assume that the field of interest is in the *far zone* of the source, which implies that $k_\perp << \beta_0$, then we can use the first-order Taylor expansion in terms of $k_\perp/\beta_0$, namely $1/(\beta_0^2 - k_\perp^2) \approx (1 + k_\perp^2/\beta_0^2)/\beta_0^2$. Finally, the finite size of the source limits the spatial frequency bandwidth in the far field by setting a maximum value of $\mathbf{k}_\perp$, say $\mathbf{k}_M$ (see figure 8.11), such that $\int_{A_{k_\perp}} d^2 \mathbf{k}_\perp = \int_{A_{k_\perp}} 2\pi k dk = \pi k_M^2$. Under these circumstances, equation (8.71) simplifies to

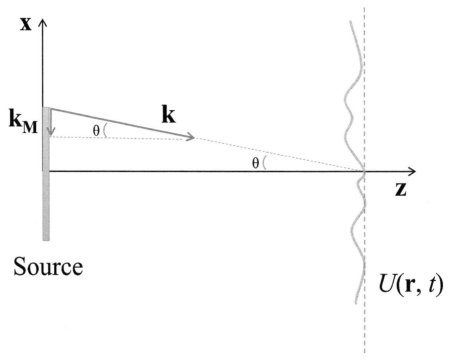

Figure 8.11. Field propagation from an extended source. At the observation plane, field U contains the maximum spatial frequency, $\mathbf{k}_M$, which is set by the half-angle θ subtended by the source.

$$\langle k_\perp^2 \rangle = \beta_0^2 \left(1 - \frac{1}{1 + k_M^2/2\beta_0^2}\right) \tag{8.72}$$

$$\approx k_M^2/2$$

where we employed a Taylor expansion a second time. The maximum transverse wavevector can be expressed in terms of the half-angle subtended by the source, θ, because $k_M = \beta_0 \sin\theta$. Thus, the coherence area of the observed field is

$$A_c = 1/\langle k_\perp^2 \rangle$$
$$= \frac{2\,\lambda^2}{\pi\,\Omega}, \tag{8.73}$$

where Ω is the solid angle subtended by the source from the plane of observation, $\Omega = 4\pi \sin^2\theta$. Note that this result matches what we obtained earlier in section 8.2. by applying Fraunhofer diffraction (equation (8.29)).

This simple calculation captures the power of using deterministic signals associated with random fields as a means to reduce the coherence propagation equation to the regular, deterministic wave equation. Specifically, by taking the power spectrum of the solution, we were able to directly calculate the second moment of the transverse wavevector and implicitly obtain an expression for the spatial coherence of the propagating field. Equation (8.73) illustrates the remarkable result that, upon propagation, the field gains spatial coherence. In other words, the free-space propagation acts as a spatial *low-pass filter.* The farther the distance from the source, the smaller the solid angle Ω and, thus, the larger the coherence area.

As a final note, let us consider the use of the deterministic signal associated with a random field in the context of far-zone (Fraunhofer) diffraction. For the deterministic fields, the following relationship applies between the source field and that at the plane z (equation (8.12))

$$V_U(x, y, z) = A V_U(k_x, k_y, 0; \omega)\Big|_{\substack{k_x = \frac{\beta_0 x}{z} \\ k_y = \frac{\beta_0 y}{z}}}, \tag{8.74}$$

where the field at $z = 0$ is just the source field $V_U(k_x, k_y, 0; \omega) = V_s(k_x, k_y, 0; \omega)$. The spatial correlation of V_U at plane z can be expressed in terms of the Fourier transform of a power spectrum, namely

$$W(\boldsymbol{\rho}, z; \omega) = V_U(x, y, z) \otimes_{x,y} V_U(x, y, z)$$
$$= \Im\left\{\left| V_U(k_x, k_y, z; \omega)\right|^2\right\}$$
$$= \Im\left\{\left| V_U(x, y, 0; \omega)\right|^2\right\} \tag{8.75}$$
$$= \Im\left\{I_U(x, y, 0; \omega)\right\}$$

Equation (8.75) shows that the spatial autocorrelation function at plane z is the Fourier transform of the intensity districbution of the source field. This result is the *far-zone* form of the van Cittert–Zernike approximation and can be very useful in calculating coherence properties of fields far away from an extended source.

8.9 Problems

1. An optical field consists of a superposition of light from two LEDs, each of Gaussian power spectra, of mean angular frequency $\omega_{1,2}$ and standard deviations $\Delta\omega_{1,2}$. Calculate the coherence time of the field.
2. Two stars subtend each an angle of 0.1 arc second as seen from Earth. The angular distance between them is 1 arc second. Calculate the coherence area for a wavelength of 500 nm.
3. An optical signal has a rectangular spectrum amplitude of width $\Delta\omega = \omega_{max}-\omega_{min}$. What is the shortest pulse that it can generate?
4. A lens of numerical aperture NA and focal distance f captures light from a circular aperture, of uniform amplitude. What is the smallest focal spot that the lens can generate at its back focal plane?
5. A lens of numerical aperture NA and focal distance f captures light from a circular aperture or radius a, of uniform amplitude. What is the coherence area at the back focal plane?
6. Redo problem 5 when the aperture is in the shape of an infinitely thin ring, of radius a.
7. Redo problem 6 for a finite ring or radius a and thickness b.

References

[1] Popescu G 2018 *Principles of Biophotonics, Volume 1—Linear Systems and the Fourier Transform in Optics'* (Bristol: IOP Publishing)
[2] Popescu G *Principles of Biophotonics, Volume 5—Light Propagation in Dispersive Media'* (Bristol: IOP Publishing) (not yet published)
[3] Popescu G *Principles of Biophotonics, Volume 9—Optical Imaging* (Bristol: IOP Publishing) (not yet published)
[4] Popescu G 2019 *Principles of Biophotonics, Volume 2—Light Emission, Detection, and Statistics* (Bristol: IOP Publishing)
[5] Glauber R J 1963 The quantum theory of optical coherence *Phys. Rev.* **130** 2529
[6] Abbe E 1873 Beiträge zur Theorie des Mikroskops und der mikroskopischen Wahrnehmung *Arch Mikrosk Anat* **9** 431
[7] Popescu G 2011 *Quantitative Phase Imaging of Cells and Tissues* (New York: McGraw-Hill)
[8] Streibl N 1985 Three-dimensional imaging by a microscope *J. Opt. Soc. Am. A* **2** 121–7
[9] Wolf E 2003 Correlation-induced changes in the degree of polarization, the degree of coherence, and the spectrum of random electromagnetic beams on propagation *Opt. Lett.* **28** 1078–80
[10] Nguyen T H, Edwards C, Goddard L L and Popescu G 2014 Quantitative phase imaging with partially coherent illumination *Opt. Lett.* **39** 5511–4
[11] Born M and Wolf E 1999 *Principles of Optics: Electromagnetic Theory of Propagation, Interference and Diffraction of Light.* 7th expanded edn (Cambridge; New York: Cambridge University Press) xxxiii 952

[12] Wolf E 1982 New theory of partial coherence in the space-frequency domain. 1. Spectra and cross spectra of steady-state sources *J. Opt. Soc. Am.* **72** 343–51

[13] Mandel L and Wolf E 1995 *Optical Coherence and Quantum Optics.* (Cambridge; New York: Cambridge University Press) xxvi 1166

[14] Kim T, Zhu R, Nguyen T H, Zhou R, Edwards C, Goddard L L and Popescu G 2013 Deterministic signal associated with a random field *Opt. Express* **21** 20806–20

[15] Nguyen T H, Majeed H and Popescu G 2015 Plane-wave decomposition of spatially random fields *Opt. Lett.* **40** 1394–7

[16] Langevin P 1908 On the theory of Brownian motion *C R Acad Sci (Paris).* **146** 530

[17] van Cittert P H 1934 Die wahrscheinliche Schwingungsverteilung in einer von einer Lichtquelle direkt oder mittels einer Linse beleuchteten Ebene *Physica* **1** 201–10

[18] Zernike F 1938 The concept of degree of coherence and its application to optical problems *Physica* **5** 785–95

[19] Goodman J W 2000 *Statistical Optics* (Wiley Classics Library Edition) (New York: Wiley) xvii 550

CPSIA information can be obtained
at www.ICGtesting.com
Printed in the USA
BVHW091402230123
656820BV00004B/86

9 780750 316453